Accelerated Reliability Engineering

WILEY SERIES IN QUALITY AND RELIABILITY ENGINEERING

Editor:
Patrick D.T. O'Connor

Electronic Component Reliability:
Fundamentals, Modelling, Evaluation and Assurance
Finn Jensen

Integrated Circuit Failure Analysis:
A Guide to Preparation Techniques
Friedrich Beck

Measurement and Calibration Requirements
for Quality Assurance to ISO 9000
Alan S. Morris

Accelerated Reliability Engineering:
HALT and HASS
Gregg K. Hobbs

Accelerated Reliability Engineering:

HALT and HASS

Gregg K. Hobbs
Hobbs Engineering Corporation, Westminster, Colorado, USA

JOHN WILEY & SONS LTD.
Chichester • New York • Weinheim • Brisbane • Singapore • Toronto

Copyright © 2000 by John Wiley & Sons Ltd,
Baffins Lane, Chichester,
West Sussex PO19 1UD, England

National 01243 779777
International (+44) 1243 779777
e-mail (for orders and customer service enquiries): cs-books@wiley.co.uk
Visit our Home Page on http://www.wiley.co.uk or http://www.wiley.com

Reprinted June 2001, September 2002

All Rights Reserved. No part of this publication may be reproduced, stored in a retrieval system, or transmitted, in any form or by any means, electronic, mechanical, photocopying, recording, scanning or otherwise, except under the terms of the Copyright, Designs and Patents Act 1988 or under the terms of a licence issued by the Copyright Licensing Agency Ltd, 90 Tottenham Court Road, London, W1P 0LP, UK, without the permission in writing of the Publisher.

Neither the author(s) nor John Wiley & Sons Ltd accept any responsibility or liability for loss or damage occasioned to any person or property through using the material, instructions, methods or ideas contained herein, or acting or refraining from acting as a result of such use. The author(s) and Publisher expressly disclaim all implied warranties, including merchantability or fitness for any particular purpose. There will be no duty on the author(s) or Publisher to correct any errors or defects in the software.

Designations used by companies to distinguish their products are often claimed as trademarks. In all instances where John Wiley & Sons is aware of a claim, the product names appear in initial capital or all capital letters. Readers, however, should contact the appropriate companies for more complete information regarding trademarks and registration.

Other Wiley Editorial Offices

John Wiley & Sons, Inc., 605 Third Avenue,
New York, NY 10158-0012, USA

WILEY-VCH Verlag GmbH
Pappelallee 3, D-69469 Weinheim, Germany

John Wiley & Sons Australia Ltd, 33 Park Road, Milton,
Queensland 4064, Australia

John Wiley & Sons (Asia) Pte Ltd, 2 Clementi Loop #02-01,
Jin Xing Distripark, Singapore 129809

John Wiley & Sons (Canada) Ltd, 22 Worcester Road,
Rexdale, Ontario M9W 1L1, Canada

Library of Congress Cataloging-in-Publication Data

Hobbs, Gregg K.
   Accelerated reliability engineering : HALT and HASS / Gregg K. Hobbs.
     p. cm. — (Wiley series in quality and reliability engineering)
   Includes bibliographical references.
   ISBN 0-471-97966-X (cased : alk. paper)
   1. Accelerated life testing. 2. Strains and stresses.
   3. Reliability (Engineering) I. Title. II. Series.
   TA169.3.H63   2000
   620'.00452—dc21
                                                    99-37601
                                                       CIP

British Library Cataloguing in Publication Data

A catalogue record for this book is available from the British Library

ISBN 0-471-97966-X

Typeset in 11.5/14pt Palatino by Footnote Graphics, Warminster, Wiltshire
Printed and bound by Antony Rowe Ltd, Eastbourne

To Virginia

Contents

Preface		xi
Series Foreword		xvii
About the Author		xxi

CHAPTER 1 Introduction to HALT and HASS: The New Quality and Reliability Paradigm — 1

- 1.1 Introduction — 1
- 1.2 Why things fail — 2
    - 1.2.1 The bathtub curve — 3
- 1.3 A short overview of HALT and HASS — 4
- 1.4 The purposes of HALT and HASS — 8
- 1.5 An historical review of screening — 12
- 1.6 The phenomenon involved — 15
- 1.7 Equipment required — 17
- 1.8 Examples of successes from HALT — 19
- 1.9 Some general comments on HALT and HASS — 23
- 1.10 Summary — 26
- References and notes — 27

CHAPTER 2 HALT: Highly Accelerated Life Tests — 31

- 2.1 Introduction — 31
- 2.2 Definition of terms — 34
- 2.3 Stimuli applied in HALT — 35
    - 2.3.1 Temperature — 37
    - 2.3.2 Vibration — 46
    - 2.3.3 Voltage — 47
    - 2.3.4 Four corner tests — 48
    - 2.3.5 Other stresses — 49
- 2.4 The general approach to HALT — 49
    - 2.4.1 The protection of fragile elements — 51

	2.5	Is one HALT enough?	53
		2.5.1 Suggested step stress intervals	53
		2.5.2 Where to stop step stress and ruggedization	54
	2.6	At what level should we do HALT?	56
	2.7	Who should do it?	57
	2.8	Where should HALT be done?	58
	2.9	How many units should we HALT and for what else can we use them?	60
	2.10	Repeated HALT (re-HALT)	62
		2.10.1 Verifying design capability	62
		2.10.2 Verifying margins	63
	2.11	Why a paradigm change is required to do successful HALT	64
		2.11.1 Failure modes and their relation to stress	65
	2.12	Time and cost savings from HALT	66
	2.13	Becoming "product smart"	69
	2.14	The biggest resistances to accepting HALT	70
	2.15	MTBF	72
	2.16	Summary	73
		References	75

CHAPTER 3 HASS: Highly Accelerated Stress Screens   77

3.1	Introduction	77
	3.1.1 Precipitation and detection screens	78
3.2	The objectives of HASS	80
3.3	Selecting the stress magnitudes	81
3.4	An example of the results of HASS alone	82
3.5	Screening effect on reliability	83
3.6	Product response to the stimuli	84
3.7	Selecting of precipitation screens	85
3.8	Selection of detection screens	87
3.9	Modulated excitations	90
3.10	Discriminators	93
3.11	Setting up the HASS regimen	95
3.12	Safety of HASS	96
3.13	The effectiveness of HASS	98
3.14	Fault coverage and resolution	99
3.15	HASS on field returns	99
3.16	Conclusions	100
	References	102

CHAPTER 4 Proof of HASS   103

4.1	Introduction	103
4.2	Safety of HASS	104

	4.3	Proof of the effectiveness of HASS	107
	4.4	Summary	112

CHAPTER 5   HASS Optimization   113

5.1	Introduction	113
5.2	The HASS optimization process	114
5.3	Conclusions on HASS	118
5.4	Highly Accelerated Stress Audit (HASA)	119
	Reference	120

CHAPTER 6   Uniformity and Repeatability   121

6.1	Introduction	121
6.2	Uniformity	121
6.3	Repeatability	125
6.4	Conclusions	126
	References	127

CHAPTER 7   Physics of Failure   129

7.1	Introduction	129
7.2	Mechanical fatigue damage – how HALT and HASS work	131
7.3	Example of vibration	134
7.4	Example of rate of change of temperature	135
7.5	What stress to use	137
7.6	Venn diagrams	140
7.7	Conclusions	143
	References and notes	143

CHAPTER 8   Software HALT: Accelerated Software Coverage and Resolution   147

8.1	Introduction	147
8.2	Detection	148
8.3	Coverage	149
8.4	Resolution	150
8.5	Solving the dilemma	151
8.6	Software fault insertion	152
8.7	Simulated field experience	152
8.8	Field experience with customer use	152
8.9	Manual fault injection	152
8.10	Automated fault injection (AFI)	153
8.11	Shipped defect level in production	157
8.12	Automated signal integrity testing	161
8.13	Time domain reflectometry (TDR)	162

x CONTENTS

	8.14	Conclusions on software HALT and signal integrity testing	162
		References	163

CHAPTER 9 **Equipment Used for HALT and HASS** 165
    9.1 Introduction 165
    9.2 Vibration equipment 166
    9.3 Various vibration systems 167
    9.4 Thermal equipment 174
    9.5 Distributed excitation 179
    9.6 Getting started 182
    9.7 Automated HASS 182
    9.8 Summary and conclusions 183
          References 184

CHAPTER 10 **Management Aspects of HALT and HASS** 185
    10.1 Introduction 185
    10.2 Proof of concept 185
          10.2.1 Training: the optimum way 186
          10.2.2 Trial in-house with new equipment: the suboptimum way 188
          10.2.3 Calculated savings: the financial estimate approach 189
          10.2.4 Trial in-house with existing equipment: use what you have 190
          10.2.5 Trial out-of-house: use an outside laboratory 191
          10.2.6 Conclusions of proof of concept 193
    10.3 Other uses of HALT and HASS 194
          10.3.1 Venture capital funding 194
          10.3.2 Acquisition of a company and due diligence 195
          10.3.3 Evaluation of competing vendors 195
    10.4 Evaluation of other HALT, HASS and HASA programs 197
          10.4.1 HALT 197
          10.4.2 HASS and HASA 199
    10.5 Outsourcing 200
    10.6 Pitfalls to avoid in HALT and HASS 202
          10.6.1 Common mistakes 202
    10.7 Using MIL-HDBK 217 or its derivatives 209
    10.8 Summary 210
          References and notes 212

APPENDIX 1 Glossary of Terms 213
Index 225

Preface

The establishment of a new paradigm has been long and arduous. A quote from a classic play comes to mind:

'There is nothing more difficult to take in hand, more perilous to conduct, or more uncertain in its success, than to take the lead in the introduction of a new order of things.' Nicolo Machiavelli, 1513, from *The Prince*.

Highly Accelerated Life Tests (HALT) and Highly Accelerated Stress Screens (HASS) techniques are accelerated methods which use stresses higher than the field environments to expose and then improve design and process weaknesses, respectively. The methods have been in use by the author and many of the author's consulting clients for three decades as of the writing of this book. Even now, in 1999, these methods are still closely guarded secrets by many of the most advanced companies using them. These companies do not publish results because to do so would be to give away to their competitors the pronounced financial and technical advantages of the techniques over the classical methods. There simply is little comparison in the speed, accuracy and cost of the classical approaches compared with the accelerated approaches which achieve:

1. time compression factors in the millions;
2. Return on Investment (ROI) factors ranging up to millions in many cases; and
3. reliability improvement factors ranging in the thousands.

HALT and HASS are proactive techniques to improve products, not to measure the reliability of the product as is attempted in

MTBF testing. This book presents the philosophy and the methodology as practised by the originator of the methods, the author. The methods are proactive in that one seeks out weaknesses in the product using extreme overstress conditions in HALT prior to release to production and then uses overstress conditions in HASS during production to quickly expose process flaws. The methods are universally successful if applied consistently and completely. Without all of the steps described herein, efforts may well be less than worthless; that is, may actually cause a dramatic increase in field failures as has been observed by the author on several occasions when some critical step has been omitted, resulting in many field failures during the warranty period or afterwards. Please note that many practitioners in industry today purport to be doing HALT and/or HASS, but are not doing them as described herein and therefore are not having the successes that are possible when the techniques are correctly applied. HALT and HASS are not new specifications on how to test reliability; they are techniques on how to actively seek to design and manufacture flaw-free and failure free products in a very cost-effective manner. They truly are methods to make products which virtually never fail.

It is readily seen that methods to produce products which never fail and keep the customer happy are a real opportunity for profit improvement. Conversely, products which fail and make the customer unhappy will severely hurt profits as seen in the results of studies done by Technical Research Programs, Washington, DC, which show that:

- In the average business, for every customer who bothers to complain, there are 26 others who remain silent.

- The average "wronged" customer will tell 8 to 16 people.

- Ninety-one percent of unhappy customers will never purchase again from the offending company.

- If an effort is made to remedy the customer's complaints, then about 90% will stay with the company.

- It costs five times as much to attract a new customer as it costs to keep an old one.

This text on HALT and HASS is based on seminars by the author on the subject. These seminars have evolved as the techniques and equipment to perform them have improved over the years. The first commercial seminars were given in 1981. As this book is written, there are two seminars, one is a two-day lecture seminar and the second is a one-day hands-on laboratory exercise to illustrate the methods using appropriate equipment. This book is intended to be a complete coverage of the HALT and HASS methods and philosophy; however, subjects needed for correct application of the methods but available elsewhere will not be covered. Among these subjects are stress analysis, dynamic analysis and the physics of failure. Any of these are the subject of several texts on the one subject alone. In these cases, references as to where to find the necessary information will be given.

The order of presentation will allow those familiar with the physics of failure and equipment of the types used for HALT and HASS to read the book in the presented order with good comprehension. Those not familiar with the subjects may want to skip ahead and read Chapter 7 on the physics of failure and Chapter 10 on equipment before reading Chapter 2 on HALT. Chapter 1 is intended to provide a basis for the more detailed discussions to follow in later chapters. It is important to note that the methods are still rapidly evolving, as is the equipment required to perform the techniques. The introductory chapter is an overview of the methods and also gives a chronology of their development. Some of the products successfully improved by the methods are listed at the end of the chapter. There are many other products on which HALT and HASS have been successfully performed, but non-disclosure agreements preclude publication of even the products. Chapter 1 touches very briefly on the physics of failure so that some appreciation of what it is all about can be gained. Chapter 2 delves into the technique of HALT; that is, how to find design and process defects very quickly using accelerated stress methodology during the design phase of the project. Note that detection screens and modulated excitation are usually necessary in order to detect the flaws exposed in HALT. These subjects are covered in Chapter 3, where the general subject of HASS is addressed and the methods for setting up a HASS profile using the information learned about the product in HALT. Once a profile has been selected for HASS, it is prudent to the point of being mandatory to demonstrate that the profile leaves sufficient life in the product, which is covered in Chapter 4. Also covered is proof of HASS effectiveness to show that the screens will indeed expose the flaws intended. Chapter 5 presents how to optimize HASS, which also generally assures a minimum cost profile while also resulting in a very effective one.

Strict uniformity and repeatability are generally not necessary in HASS and Chapter 6 gives the reasons why this is true. A careful study of the methods described in Chapter 6 has saved many companies substantial sums of money and has provided a true acceleration of the whole process from design through to full scale manufacturing by allowing less than perfect uniformity of stresses applied during both HALT and HASS. The reasons why time compression works and, therefore, why HALT and HASS work are covered in Chapter 7 on the physics of failure. Chapter 8 delves into the subjects of coverage and resolution in the test systems used to determine that the product is properly designed and assembled and is named Software HALT since it enhances the speed at which the software is brought to the mature state with good coverage and resolution. This is an activity apart from the usual HALT and is done as a software development tool. Lack of coverage has been found to be the cause of many "No defects found", "Cannot duplicate" or "Could not verify" results from field failures. Major progress is underway in this field, but again the leaders are not publishing because of the great technological and cost advantages of using the methods. Without good coverage, HALT and HASS will be significantly reduced in effectiveness; that is, if one cannot detect that a problem exists, then one cannot address the problem.

Equipment for the performance of HALT and HASS are discussed in Chapter 9, and some reasons for the use of the equipment specifically designed to perform HALT and HASS are also discussed. Proof of Concept, that is to demonstrate that HALT and HASS work on a specific product line, is discussed in Chapter 10. There are many other uses of HALT and HASS, which may not be obvious to the novice, that have been utilized and a few of them are also discussed in Chapter 10. It is frequently necessary to evaluate the programs of suppliers or perhaps a company being considered for acquisition and Chapter 10 covers this subject, including the impacts of outsourcing on cost, quality and reliability. Additionally, Chapter 10 also summarizes subjects covered in the book, draws conclusions and discusses some common mistakes observed by the author. Reference to this chapter is excellent advice for current practitioners who have not studied the techniques as presented in this book. A glossary of terms used in this book is given in Chapter 11.

It is noted that many published papers on HALT and HASS have serious omissions in technique and/or equipment used and, therefore, the results could be expected to be poor at best, if not completely counterproductive. Several major disasters have occurred because someone not fully trained in the methodology attempted to use HALT and HASS and omitted some critical step. *The methods are completely safe if and only if used consistently and correctly.*

Technical materials readily available in other texts have not been included for the sake of brevity. Numerous references are given for the purpose of completeness.

The reading and comprehension of the entire text is necessary for a thorough understanding of the methodology of HALT and HASS. It is suggested that anyone who intends to participate in any significant way in a HALT and HASS program should become not only familiar with the entire text, but to be expert in all aspects of it as well. The results of a properly done HALT and HASS program are truly rewarding on both the technical and the financial levels.

Enjoy your successes using HALT and HASS, I certainly have.

Gregg K. Hobbs
Westminster, Colorado, USA
April 1999

Series Foreword

Modern engineering products, from individual components to large systems, must be designed and manufactured to be reliable in use. The manufacturing processes must be performed correctly, and with the minimum of variation. All of these aspects impact upon the costs of design, development, manufacture and use, or, as they are often called, the product's life cycle costs. The challenge of modern competitive engineering is to ensure that life cycle costs are minimized, whilst achieving requirements for performance and time to market.

If the market for the product is competitive, improved quality and reliability can generate very strong competitive advantages. We have seen the results of this in the way that many products, particularly Japanese cars, machine tools, earthmoving equipment, electronic components and consumer electronic products have won dominant positions in world markets in the last 30 to 40 years. Their domination has been largely the result of the teaching of the late W. Edwards Deming, who taught the fundamental connections between quality, productivity and competitiveness. Today this message is well understood by nearly all engineering companies that face the new competition, and those that do not understand lose position or fail.

Concurrently with the philosophy and methods that took root initially in Japan and then spread back to the West where most originated, methods were developed in the USA to address the problems of unsatisfactory quality and reliability of military equipment. These included formal systems for quality and reliability management (MIL-Q-9858 and MIL-STD-758) and methods for predicting and measuring reliability (MIL-STD-721, MIL-HDBK-217, MIL-STD-781), MIL-Q-9858 was the model for the international standard on quality systems (ISO 9000), and the methods for quantifying reliability have been similarly developed and applied to other types of product.

The methods developed in the West were driven to a large extent by the customers, particularly the military. They reacted to perceived low achievement by the imposition of standards and procedures, whilst their suppliers saw little motivation to improve, since they were paid for spares and repairs. By contrast, the Japanese quality movement was led by industry, who had learned how quality provided the key to greatly increased productivity and competitiveness.

These two streams of development epitomize the difference between the deductive mentality applied by the Japanese to industry in general, and to engineering in particular, in contrast to the more inductive Western approach. The deductive approach seeks to generate continuous improvements across a broad front, and new ideas are subjected to careful evaluation. The inductive approach leads to inventions and 'break-throughs', and to greater reliance on 'systems' for control of people and processes. The deductive approach allows a clearer view, particularly in discriminating between sense and nonsense. However, it is not conducive to the development of radical new ideas. Obviously these traits are not exclusive, and most engineering work involved elements of both. However, the overall tendency of Japanese thinking shows up in their enthusiasm and success in industrial teamwork and in the way that they have adopted the philosophies of Western teachers such as Deming and Drucker, whilst their Western competitors have found it more difficult to break away from the mould of 'scientific' management, with its reliance on systems and more rigid organizations and procedures.

Unfortunately, the development of quality and reliability engineering has been afflicted with more nonsense than any other branch of engineering. This has been the result of the development of methods and systems for analysis and control that contravene the deductive logic that quality and reliability are achieved by knowledge, attention to detail, and continuous improvement on the part of the people involved. Of course Western minds have also made great positive contributions: we need only recall Shewhart's invention of statistical process control and Fisher's invention of statistical experiments, and of course Deming was an American. Therefore it can be difficult for students, teachers, engineers and managers to discriminate effectively, and many have been led down wrong paths.

In this series we will attempt to provide a balanced and practical source covering all aspects of quality and reliability engineering and management, related to present and future conditions, and to the range of new scientific and engineering developments that will shape future products. I hope that the series will make a positive contribution to the teaching and practice of engineering.

Patrick D.T. O'Connor
August 1994

About the Author

Gregg K. Hobbs was born in Hollywood, California, in 1939, lived in California and then attended UCLA from 1960 until 1965. He completed his BS degree in Engineering in 1963. In 1965 he left the university after completing all course work for the Ph.D. program with a major in fluid mechanics and minors in structural dynamics, strength of materials and applied mathematics. He then worked in industry at TRW Systems Corporation in Redondo Beach, California, and at Hughes Space System Division in Los Angeles, California. While at Hughes he obtained his MSME from UCLA by examination. He then moved to Santa Barbara, California, in 1967 and worked at Astro Research Corporation designing and testing deployable space structures until 1969 when he returned to graduate school at UCSB. He completed the Ph.D. program in 1970 with a major in structural dynamics and with minors in non-linear analysis, control systems and strength of materials. Upon completion of his technical training, he obtained Professional Engineer registration in Mechanical Engineering, Control Systems Engineering and in Civil Engineering, all in California.

Dr. Hobbs was employed by Santa Barbara Research Center from 1970 until 1976 as head of the Dynamics and Stress Analysis Group. Moving to Minneapolis, Minnesota, he was employed by MTS Systems Corporation, a manufacturer of hydraulic test equipment, as a Senior Applications Engineer. In 1978, he founded Hobbs Engineering Corporation where he was President and held seminars and acted as consultant in the United States, Europe, Canada and the near East. In 1990, he was the sole founder of the QualMark Corporation which manufacturers equipment for the performance of HALT and HASS as well as operating laboratories to perform the process on a contract basis. He functioned as President, Chairman of the Board of Directors and Chief Technical Officer. In 1995 he returned to work full time in Hobbs Engineering Corporation.

Technical articles published by him number over 100 and cover several technical fields. As of 1999, Dr Hobbs has 9 US and many foreign patents to his credit and has several more pending and in preparation. He is a prolific inventor in several technical fields including structural dynamics, pneumatics, optics, acoustics, surgical instruments, gas dynamics and thermodynamics. His interests include flying and he has a private license with multi-engine and instrument ratings. He has his own multi-engine airplane, which he flies on business when possible. His home is in Westminster, Colorado, where he lives with his wife, Virginia.

CHRONOLOGY OF HALT AND HASS ACTIVITIES BY THE AUTHOR

1965 First attempt to do Design Ruggedization met with rejection.

1966 Ruggedized design analysis of the Lunar Module rejected.

1967 First completed Design Ruggedization on a Cassegranian telescope structure.

1968 Ruggedized design of a Navy electro-optics system.

1969 Ruggedized design of a space probe radiation cooler.

1970 Ruggedized design of the Multi-Spectral Scanning Instrument for the Earth Resources Technology Satellite, NASA.

1978 Founded Hobbs Engineering Corporation to offer consultations and run seminars.

1979 Ruggedized Design applied to a US Air Force Sidewinder missile proximity fuse.

1979 First use of Enhanced Environmental Stress Screening (ESS).

1979 Proof of Screen conceived and utilized.

1980 Tickle Vibration used in detection screens.

1980 Screen Optimization put into practice.

1980 Seminars on Design Ruggedization and Enhanced ESS offered publicly.

1983 Design of a flexible six-axis vibration system.

1984 Design of a relatively rigid six-axis vibration system.

1988 HALT coined to differentiate from Design Ruggedization.

1988 HASS coined to differentiate from Enhanced ESS.

1989 Precipitation and Detection Screens introduced.

1990 Founded QualMark Corporation to build equipment for HALT and HASS.

1996 Modulated Excitation conceived and put into practice.

1996 Design of the six-axis Modular™ Vibration System.

1997 Design of the six-axis Broad Band™ Vibration System.

1998 Software HALT coined and published.

1999 Energy redistribution system conceived.

CHAPTER 1

Introduction to Halt and Hass: The new quality and reliability paradigm

'Great spirits have always encountered violent opposition from mediocre minds.'

Albert Einstein

1.1 INTRODUCTION

Highly Accelerated Life Tests (HALT) and Highly Accelerated Stress Screens (HASS) are briefly introduced and discussed. A few successes from each technique are described. These techniques have been successfully used by the author and some of the author's consulting clients and seminar attendees for three decades. Most of these users do not publish their results because of the pronounced financial and technical advantages of the techniques over the classical methods, which are not even in the same league in terms of speed and cost. Later chapters will cover the methods in much greater detail and this introduction is intended to provide a basis for the more detailed discussions to follow in later chapters. It is important to note that the methods are still rapidly evolving as is the equipment required in order to implement the techniques. This chapter is an overview of the methods and also gives a chronology of their development.

The HALT and HASS methods are designed to improve the

reliability of the products, not to determine what the reliability is. The approach is therefore proactive as compared with a Reliability Demonstration (Rel-Demo) or Mean time between failures (MTBF) tests that do not improve the product at all but simply (attempt to) measure what the reliability is. This is a major difference between the classical and the HALT approaches.

1.2 WHY THINGS FAIL

A product will fail when the applied load exceeds the strength of the product. The load can be voltage, current, force, temperature or other variable. Consider applied load and strength plotted together as in Figure 1.1.

Whenever the applied load exceeds the strength, failure will occur. The load may be a one time load or it may be applied a number of times. In the first case, overload failure will occur and in the second case fatigue failure will occur. A figure could be drawn for either case and would look similar to that shown. The cross-hatched area represents the products which will fail.

As time passes, the product could become weaker for any one of many reasons. Figure 1.2 is concerned with aging. Alternatively, one could depict fatigue damage by having the strength curve move to the left as depicted in Figure 1.2. Again, when the applied load exceeds the strength, failure will occur. Either way, the overlap of the curves will increase, meaning that more products will fail. This moving of the curve can also be simulated by moving the applied load curve to the right as depicted in Figure 1.3. Note that one would have the same failures as when the strength degraded. It is this last approach that is taken in HALT, wherein the loads are

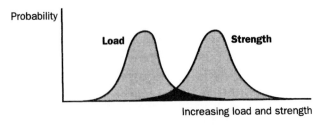

Figure 1.1 Load and strength

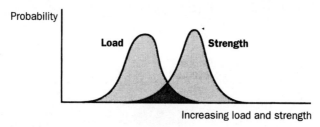

Figure 1.2

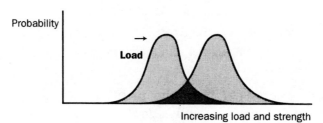

Figure 1.3

increased until failure occurs, identifying a weakness. It is seen that one would obtain the same failures in either case according to the illustration. This simplistic example is quite valid and one could go through detailed calculations to demonstrate the fact. It will be left as a simple illustration here.

1.2.1 The Bathtub Curve

The pattern of failures that occurs in the field can be approximated in three ways. When there are defects in the product, so-called "infant mortalities", or failure of weak items, will occur. Another type of failure is due to externally induced failures where load exceeds strength. Finally, wearout will occur even if an item is not defective. When one superimposes all three types of failure, a curve called the bathtub curve occurs. One such curve is shown in Figure 1.4.

The bathtub curve is grossly affected by HALT and HASS techniques:

1. Production screening (HASS), will reduce the early segment of the curve by eliminating early life failures due to manufacturing flaws.

4 INTRODUCTION TO HALT AND HASS

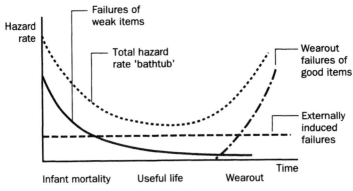

Figure 1.4 The bathtub curve

2. Ruggedization (HALT) of the product will lower the mid-portion of the curve which is due to externally induced failures.

3. HALT will extend the wearout segment far to the right.

All of these will be discussed at length later in the book.

1.3 A SHORT OVERVIEW OF HALT AND HASS

HALT is an acronym for Highly Accelerated Life Test, which was coined by the author in 1988 after having used the term "Design Ruggedization" for several years. The term HALT is not trademarked when used to describe techniques, nor is it service-marked when used to describe services. You are encouraged to use the term if you are doing HALT as described in this book because it explicitly describes the methodology and is used worldwide. The same applies to HASS. In HALT, every stimulus of potential value is used under accelerated test conditions during the design phase of a product in order to find the weak links in the design and fabrication processes. Each weak link found provides an opportunity to improve the design or the processes, which will lead to reduced design time, increased reliability and decreased costs. HALT compresses the design time and therefore allows earlier (mature) product introduction. Studies have shown that a six-month advantage in product introduction can result in a lifetime profit increase of 50% [2].

The stresses applied in HALT and HASS are not meant to simulate the field environments at all, but are meant to expose the weak links in the design and processes using only a few units and in a very short period of time. The stresses are stepped up to well beyond the expected field environments until the "fundamental limit of the technology" ([1] and [3] and Chapter 2) is reached in robustness. Reaching the fundamental limit generally requires improving everything relevant found, even if found at above the "qualification" levels! This means that one ruggedizes the product as much as possible without unduly spending money – sometimes called "gold plating". Only those failures that are likely to occur in normal environments are addressed. Intelligence must be used. One could easily overdo the situation and make the product much more rugged than necessary and spend unnecessary money in the process. Most of the weaknesses found in HALT are simple in nature and inexpensive to fix, such as:

1. A capacitor which flies off the board during vibration can be bonded to the board or moved to a node of the mode causing the problem

2. A particular component which is found to have a less than optimum value can be changed in value.

3. A screw backed out during vibration and thermal cycling can be held in with a thread locker.

HALT, or its predecessor, Design Ruggedization, has, on many occasions since 1969, provided substantial (50 to 1000 times, or, in a few cases, even higher) reductions in field failures, time to market, warranty expenses, design and sustaining engineering time and cost and also the total development costs. One of the main benefits of HALT is that it minimizes the sample size necessary to do design verification testing. It is noted in passing that STRIFE (*STR*ESS plus L*IFE*), as used by Hewlett-Packard, is a subset of HALT. The basic philosophy of HALT has been in use since 1969 and has been used by the author and others on many products from various fields. Some of these fields are listed in Table 1.1 at the end of this chapter.

The first complete ruggedized design by the author was

performed in 1969 on his mechanical design of a Casagranian telescope for the Earth Resources Technology Satellite. The system which contained the telescope was the Multi-Spectral Scanner built by Santa Barbara Research Center in Goleta, California. The telescope was built of invar and fuzed silica, both of which have very low thermal expansion coefficients. The optics were mounted by tangent bar mounts which allowed radial expansion with very little constraining force so as not to distort the optics. The telescope was nine inches in diameter and weighed only 7lb as shown in Figure 1.5. Only adhesive bonding held the optics in place. The prototype withstood 25 GRMS single-axis random vibration for several hours, temperature cycling between $-70\,°C$ and $+125\,°C$ for many cycles and then 25 GRMS vibration at the high and low temperatures without failure. The design successfully flew several years later even though a change in the scan mirror design generated severe vibration during operation. The very light weight and intentional modal tuning, as well as the high resonant frequencies of the telescope, led to optical pointing to within one microradian during the vibration due to the scan mirror. Because this early ruggedization had been done, Hughes Aircraft was able to deliver the US $100,000,000 spacecraft on time.

Some attendees at the author's seminars have, in addition, used the techniques on thousands of products, so the *basic* methods are not new, but have not been publicized much until very recently because of the tremendous advantages in reliability and cost gained by their use. The techniques continue to be improved and in 1991 the author introduced precipitation and detection screens, decreasing screen cost by at least an order of magnitude and simultaneously increasing the effectiveness by several orders of magnitude. In 1996, the author introduced the concept of a search pattern which makes the detection of precipitated defects better by *at least* one order of magnitude, if not many orders of magnitude. This concept is discussed in Chapter 3 as modulated excitation. Software HALT was introduced in 1998 and is covered in Chapter 8. In this software development technique, one exposes lack of coverage in the test hardware/software system and then improves the test coverage. The results of Software HALT are currently being held as company private by the few users and the author has

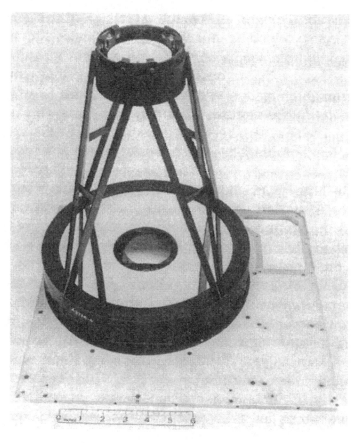

Figure 1.5 Casagranian telescope (Courtesy Santa Barbara Research Center)

seen no publications beyond those available through non-disclosure agreements with consulting clients. Paybacks measured in a few days are common in Software HALT; that is, the total cost of purchasing the test equipment and running the tests is recovered in a few days. The test equipment is then available for another series of tests and can again achieve a similar payback. This payback could be accomplished 26 times in one year (assuming a two-week payback) on this one piece of equipment. Without compounding the rate of return, the simple rate of return is 2,600% per year! This rate is so high that anyone aware of the methods would use them. Reasons such as this are why the few leaders will not publish their results and give their competition the information necessary for them to start using the same techniques.

8 INTRODUCTION TO HALT AND HASS

HASS is an acronym for Highly Accelerated Stress Screens, which was also coined by the author in 1988 after using the term "Enhanced Environmental Stress Screening (ESS)" for some years. These screens use the highest possible stresses (frequently well beyond the Qualification (QUAL) level) in order to attain time compression in the screens. Note that many stimuli exhibit an exponential relationship between stress level and "damage" done (see Chapter 7) resulting in a much shorter duration of stress (if the correct stress is used). It has been proven that HASS generates extremely large savings in screening costs since much less equipment such as shakers, chambers, consumables (power and liquid nitrogen), monitoring systems and floor space are required due to time compression in the screens. The time compression is gained in the precipitation screen (Chapter 3). The screens must be, and are proven to be, of acceptable fatigue damage accumulation or lifetime degradation using Proof of HASS techniques which are discussed in Chapter 4. Safety of HASS demonstrates that repeated screening does not degrade the field performance of the unit under test and is a crucial part of HASS development. HASS is generally not possible unless a comprehensive HALT has been performed. Without HALT, fundamental design limitations will restrict the acceptable stress levels in production screens to a very large degree and will prevent the large accelerations of flaw precipitation, or time compression, which are possible with a very robust product. It is noted that a less than robust product probably cannot be effectively screened by the "classical" screens without a substantial reduction in its field life. Chapter 5 covers this and other relevant subjects regarding the effect of screening on the field lifetime left in a product.

1.4 THE PURPOSES OF HALT AND HASS

The general purposes of applying accelerated stress conditions in the design phases is to find and improve upon design and process weaknesses in the least amount of time and correct the source of the weaknesses before production begins. It is generally true that robust products will exhibit much higher reliability than non-robust ones and so the ruggedization process of HALT in which large margins are obtained will generate products of high potential reliability. In order to achieve the potential, defect-free hardware must be manufactured or, at least, the defects must be found and fixed before shipment. In HASS, accelerated stresses are applied in production in order to shorten the time to failure of the defective units and therefore shorten the corrective action time and the number of units built with the same flaw. Each weakness found in HALT or in HASS represents an opportunity for improvement. The author has found that the application of accelerated stressing techniques to force rapid design maturity (HALT) results in paybacks that far exceed these from production stressing (HASS). Nonetheless, production HASS is cost effective in its own right until quality is such that a sample HASS or Highly Accelerated Stress Audit (HASA) can be put into place. The use of HASA demands excellent process control since most units will be shipped without the benefit of HASS being performed on them, and only those units in the selected sample will be screened for defects. These subjects will be covered in much more detail in following chapters.

The stresses used in HALT and HASS include, but are not restricted to, all-axis simultaneous vibration, high-rate broad-range temperature cycling, power cycling, voltage and frequency variation, humidity, and any other stress that may expose design or process problems. No attempt is made to simulate the field environment – one only seeks to find design and process flaws by any means possible. The stresses used generally far exceed the field environments in order to gain time compression; that is, shorten the time required to find any problem areas. When a weakness is discovered, only the failure mode and mechanism is of importance, the relation of the stress used to the field environment is of no consequence at all. Figure 1.6 illustrates this point. In this figure, λ is the instantaneous failure rate for a given failure mode. The two curves illustrating a thermally induced failure rate and a vibration-induced failure rate are located so that the field stresses at which failure occurs and the HALT stresses at which failure occurs are lined up vertically. It is then seen that a failure mode that would most often be exposed by temperature

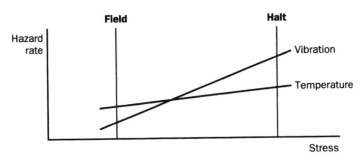

Figure 1.6 Instantaneous failure rates in the field and in HALT

will be more likely to be exposed by vibration in the HALT environment.

It is very common to expose weaknesses in HALT with a different stress than the one that would make the weakness show up in the field. It is for this reason that one should focus on the failure mode and mechanism instead of the margin for the particular stress in use.

"Mechanism" here means the conditions that caused the failure, such as melting, exceeding the stable load or exceeding the ultimate strength. The corresponding failure mode could be separation of a conductor, elastic buckling and tensile failure, respectively. Considering the margin instead of the failure mode is a major mistake which is made by most engineers used to conventional test techniques. In HALT and HASS, one uses extreme stresses for a very brief period of time in order to obtain time compression in the failures. In doing so, one may obtain the same failures as would occur in the field environments, but with a different stress. For example, a water sprinkler manufacturer had a weakness which was exposed by the diurnal thermal cycle in the field. HALT exposed the same weakness with all-axis vibration after extensive thermal cycling failed to expose the weakness. After the weakness was addressed, the field failures were eliminated, which proves that the weakness exposed by all-axis vibration was a valid discovery. For another example, consider a reduction in the cross-sectional area of a conductor. This reduction would create a mechanical stress concentration and an electrical current density concentration. This flaw might be exposed by temperature cycling or vibration in HALT or HASS and might be exposed by electromigration in the field environment. Either way, the reduction in area introduces a weakness that can be eliminated.

In addition to stresses, other parameters are used to look for weaknesses. In the author's experience, these have included the diameter of a gear, the pH of a fluid running through the product, contaminants in the fluid running through a blood analyzer, the thickness of a tape media, the viscosity of a lubricant, the size of a tube or pipe, the lateral load on a bearing and an almost endless additional number of factors. What is sought is any information that could lead to an opportunity for improvement by decreasing the sensitivity of the product to any conditions which could lead to improper performance or to catastrophic failure. Anything that could provide information for an improvement in the margin is appropriate in HALT. Accepting this philosophy is one of the most difficult shifts in thinking for many engineers not trained in the HALT–HASS approach.

In the HALT phase of product development, which should be in the early design phase, the product is improved in every way practicable bearing in mind that most of what are discovered in HALT as weaknesses will almost surely become field failures if not improved. This has been demonstrated thousands of times by users of HALT. Of course, one must always use reason in determining whether or not to improve the product when an opportunity is found and this is done by examining the failure mode and mechanism. Just because a weakness was found "out of spec" is no reason to reject the finding as an opportunity for improvement. There are numerous cases where weaknesses found "out of spec" were not addressed until field failures of the exact same type occurred. If you find it in HALT, it is probably relevant. In various papers from Hewlett-Packard over the years [3], it has been found that most of the weaknesses found in HALT and not addressed resulted in costs to the company in the neighborhood of US$10,000,000 per failure mode to address later, when failure costs were included. It cannot be emphasized too much that it is imperative to focus on the failure mode and mechanism and not the conditions used to make the weakness apparent. Focusing on the margin will usually lead one to allow a detected weakness to remain, resulting in many field failures of that type before a fix can be implemented. Learn from others' mistakes and do not focus on the stress level used, but on the failure mode and mechanism.

HALT and HASS are not restricted to electronic boxes, but apply to many other technologies as well. Some of the technologies are listed at the end of the chapter and include such diverse products as shock absorbers, lipstick, airframes, auto bodies, exhaust systems and power steering hoses to name just a few. Before getting into the techniques in general, it is beneficial to know the history of early attempts at stress screening that paralleled the development of HALT and HASS. Note that HALT addresses design and process weaknesses, whereas classical ESS only addresses production weaknesses. HASS may expose design weaknesses if any remain or are introduced after production start.

1.5 AN HISTORICAL REVIEW OF SCREENING

Many mistakes have been and currently are being made in screening and so a review of some historical events is educational in the sense that we want to go forward and learn from our collective past errors. In the 1970s, the US Navy, which was not alone in this regard, experienced very poor field service reliability. In response to this, an investigation was performed and it was found that many of the failures were due to defects in production that could be screened out using thermal cycling and random vibration. The Navy issued a stress screening guideline, NAVMAT P-9492, which laid out guidelines for production screening on Navy programs. This document was included in most Navy contracts and so production screening became a requirement on most Navy programs. Although the document was frequently treated as a MIL-SPEC, which was not the intent, tremendous gains in reliability resulted from the techniques required by the Navy. Among other things, random vibration and thermal cycling were required, but were not required to be applied simultaneously. Data acquired later by many investigators would show that combined vibration and thermal cycling is more than ten times as effective and is less expensive due to the reduction of test hardware. The author has found in many workshops with actual production hardware that no defects at all were found unless modulated excitation was used. This is covered in Chapter 3.

In the late 1970s, the Institute of Environmental Sciences (IES) began to have annual meetings which addressed the subject of ESS and then the IES issued guidelines on the production screening of assemblies in 1981 and 1984 and on the screening of parts in 1985. There were three major problems with the survey and the published results [4, 5].

1. The companies surveyed were mostly under contract to the US military and screens had been imposed by contract in many cases and had not been carefully tuned to be successful. Some had even been tuned *not* to be successful. Most of the contracts issued even made it more profitable for the contractors to produce hardware which did not have high field reliability, but would pass the screens. In these cases, the contractor could sell almost everything produced and, since the field reliability was poor, many spares were needed as was a rework facility to fix those units which failed in the field. It is readily apparent that the screens used in these types of contracts would not be the most effective screens that the technology of the day could produce.

2. The IES polled the contractors doing screening and then published the results as the "effectiveness" of various screens instead of a more accurate term of "popularity" or "what was required to be done". See [6] for some details. Among the misconceptions in the guidelines was that thermal cycling is the most effective screen. This misconception is extant to a large extent today; however, many HALT results have shown that all-axis vibration far surpasses the effectiveness of thermal cycling for the broad spectrum of faults found in many types of equipment today, including electronics. The last statement is only true if a six-axis shaker is used.

3. The guidelines emphasized 100% screening instead of emphasizing corrective action which would eventually allow the screening to be reduced to a sample. An interesting observation is that the definition of screening as a process on 100% of the production maximized the equipment, manpower and other costs! This led to a bonanza for some equipment manufacturers and for contractors working on a "cost plus fee" basis.

These three problems in the IES guidelines led many novices to try the techniques, which were relatively ineffective and inordinately expensive, features which may be a plus if one is working on a cost plus fee basis contract or if one is selling equipment to perform screening, respectively. With only the IES guidelines as a source of information, many financially and technically unsuccessful screening programs were tried. Many companies simply gave up and went back to their old ways or simply complied with contractual requirements.

In the meantime, some very successful screening techniques were developed, tried, improved and retried. These techniques are so successful compared with the guideline's methods that most of the companies using the new techniques will not publish their results, although some companies are starting to publish to a limited extent but are omitting critical factors such as design flaws eliminated, warranty return rates and returns on investments (ROIs). See Chapter 2 on HALT for some previously unpublished results and Chapter 3 on HASS for some cost savings numbers. However, from knowledge gained in consulting under non-disclosure agreements, most of the truly outstanding successes are still not being published. Many disasters from the misapplication of HALT and HASS are also not being published. Properly applied, however, HALT and HASS always work.

Along with the technique improvements, equipment for performing the new techniques was developed. When the newer equipment was available, further technique improvements were possible. This confluence of technique and equipment improvements has been in effect for several cycles. The author sees no end to the development in sight and is, at the time of the writing of this book, in the development and patent stage on equipment that will far surpass anything yet available for the techniques.

1.6 THE PHENOMENA INVOLVED

Several phenomena may be involved when screening occurs. Among these are mechanical fatigue damage, wear, electromigration, chemical reactions, as well as many others. Each of these has a different mathematical description and responds to a different stimulus or stimuli. Chemical reactions and some migration effects proceed to completion according to the Arrhenius model or some derivative of it. It is noted that many misguided screening attempts assume that the Arrhenius equation *always* applies; that is, that higher temperatures lead to higher failure rates, but this is not an accurate assumption. See Chapter 7 for more details. See [7] for many excellent discussions of the use and misuse of the Arrhenius concepts. MIL-HDBK 217, a predictive methodology for electronic reliability without any scientific basis whatsoever, was based on these concepts and therefore is quite invalid for predicting the field reliability of the products which are built today. MIL-HDBK 217 is even less valid and completely misleading when used as a reverse engineering tool to improve reliability since it will lead one to reduce temperatures even when a reduction will not reduce the failure rate and may even increase it due to changes made to decrease the temperature, such as the addition of cooling fans. That is, new failure modes may be introduced and the basic reason for some existing failures may not be changed at all. There is an excellent discussion in [8] of the temperature sensitivities of many microelectronics parts which are stated to be insensitive to temperature below 150 °C.

Many failures in electronic equipment are mechanical in nature: fatigue of a solder joint, fatigue of a component lead, failure of a pressure bond or similar modes of failure. The mechanical fatigue damage done by mechanical stresses due to temperature, rate of change of temperature, vibration, or some combination of them can be modeled in many ways, the least complex of which is Miner's Criterion. This criterion states that fatigue damage is cumulative, is non-reversible, and accumulates on a simple linear basis which in words is "The damage accumulated under each stress condition taken as a percentage of the total life expended can be summed over all stress conditions. When the sum reaches unity, the end of fatigue life has arrived and failure occurs." The data for percentage of life expended are obtained from S–N (stress level versus number of cycles to fail) diagrams for the material in question. A general relationship [9] based on Miner's Criterion follows:

$$D \approx NS^\beta, \qquad (1.1)$$

where:

D is the fatigue damage accumulated, normalized to unity,

N is the number of cycles of stress,

S is the mechanical stress (in pounds per square inch, for example), and

β is an exponent derived from the S–N diagram for the material and ranges from 8 to 12 for most materials; physically, it represents the negative inverse slope of the S–N diagram.

The flaws (design or process) that will cause field failures usually, if not almost always, cause a stress concentration to exist at the flaw location (and this is what causes the early failure). Just for illustrative purposes, let us assume that there is a stress which is twice as high at a particular location which is flawed due to an inclusion or void in a solder joint. According to the equation above with beta assumed to be about 10, the fatigue damage would accumulate about 1000 times as fast at the position with the flaw as it would at a non-flawed position having the same nominal stress level; that is, having the same applied load without the stress concentration. This means that the flawed area can fatigue and break and still leave 99.9% of the life in the non-flawed areas. Our goal in environmental stress screening is to do fatigue damage to the point of failure at the flawed areas of the unit under test as fast as possible and for the minimum cost. With the proper application of HALT, the design will have several, if not many, of the required lifetimes built into it and so an inconsequential portion of the life would be removed in HASS. This would, of course, be verified in Safety of HASS (see Chapter 4 for a complete coverage of the subject). Note that the relevant question is "How much life is left after HASS?", not "How much did we remove in HASS?" Also note that *all* screens remove life from the product. This is a fundamental fact that is frequently not understood by those unfamiliar with the correct underlying concepts of screening. A properly done HALT and HASS program will leave more than enough life remaining and will do so at a much reduced total program cost.

Flaws of other types have different equations describing the relationship between stress and the damage accumulation, but all seem to have a very large time compression factor resulting from a slight increase of the stress. This is precisely why HALT and HASS techniques work. More details are given in Chapter 7.

1.7 EQUIPMENT REQUIRED

Chapter 9 has a more complete coverage of the equipment required, but a brief introduction here is appropriate so that the reader can appreciate the book without having to read Chapter 9 first.

The application of the techniques mentioned in this book generally is very much enhanced by, if not impossible without, the use of environmental equipment of the latest design such as all-axis exciters and combined very high-rate thermal chambers (60 °C/min or more *product* rate). All-axis means three translations and three rotations.

A single-axis, single-frequency shaker will only excite modes in the particular direction of the vibration and only those nearby in frequency. A swept sine will sequentially excite all modes in the one direction being excited. A single-axis random shaker will simultaneously excite all modes in one direction. A six-axis system will simultaneously excite all modes within the bandwidth of the shaker in all directions. If all modes in all directions are not excited simultaneously, then many defects can be missed. Obviously, the all-axis shakers are superior for HALT and HASS activities since one is interested in finding as much as possible as fast as possible.

In the very early days of Design Ruggedization (the precursor to HALT), a device had been severely ruggedized using a single-axis random shaker system. This effort was reported in [10]. Then, in production, a very early all-axis system was used and three design weaknesses which had not been found on the single-axis system were exposed almost immediately. That experience showed the author the differences in the effectiveness of the various systems. Since then, the system of choice has been an all-axis broad-band shaker.

Other types of stresses or other parameters may be used in HALT. In these cases, other types of stressing equipment may be required. If one wanted to investigate the capability of a gearbox, one could use contaminated oil, out-of-specification gear sizes and a means for loading the gearbox in torsion either statically or dynamically. If one wanted to investigate various end piece crimping designs on power steering hoses, one could use temperature, vibration and oil pressure simultaneously. This has been done and worked extremely well, exposing poor designs in just a few minutes. In order to investigate an airframe for robustness in pressurization, the hull could be filled with water and rapid pressure cycling done. This is how it is done at several aircraft manufacturers. Water is used as the pressurized medium since it is nearly incompressible and so when a fracture occurs, pressure drops quickly, preventing an explosive-type failure, such as would occur if air were to be used. A life test simulating thousands of cycles can be run in just a few days using this approach.

Note that, in HALT and HASS, one tries to do fatigue damage as fast as possible, and the more rapidly it is done, the sooner it can stop and the less equipment is needed to do the job. It is not unusual to reduce equipment costs by *orders* of magnitude by using the correct stresses and accelerated techniques. This comment applies to all environmental stimulation and not just to vibration. An example discussed later in this book (Chapter 7) shows a decrease in cost from US$22 million to US$50,000 on thermal chambers alone (not counting power requirements, associated vibration equipment, monitoring equipment and personnel) by simply increasing the rate of change of temperature from 5°C/min to 40°C/min (when rate-sensitive flaws are present)! The basic data for this comparison is given in [11]. Another example shows that increasing the RMS vibration level by a factor of 1.4 times would decrease the vibration system cost from US$100 million to only US$100,000 for the same throughput of product. With these examples, it becomes clear that HALT and HASS techniques, when combined with modern screening equipment designed specifically to do HALT and HASS, provide quantum leaps in cost effectiveness.

Some typical results of HALT and HASS applied to product design and manufacturing are described below. Some of these are from early successes and have been published in some form, usually technical presentations at a company. Later examples using the later technology in terms of technique and equipment have largely not been published. The later results are, of course, much better, but the early results will make the point well enough, since they represent a lower bound on the expected successes today when far better techniques and equipment are available.

1.8 EXAMPLES OF SUCCESSES FROM HALT

1. In 1984, an electromechanical impact printer's MTBF was increased 838 times when HALT was applied. A total of 340 design and process opportunities for improvement were identified in the several HALTs which were run. All of these were implemented into the product before production began, resulting in an initial production system MTBF, *as measured in the field*, of 55 years! This product is about 10 in × 18 in × 27 in and weighs about 75 lb. It is interesting that the MTBF never got better than it was at initial product release, but it did get worse when something went out of control. The out-of-control conditions were spotted by the 5% sample HASS called Highly Accelerated Stress Audit (HASA). The reason there was no reliability growth after product introduction is that the system was born fully mature due to HALT. This is one of the major goals of HALT and it is the case *if and only if* advantage is taken of all of the discovered opportunities for improvement.

2. A power supply which had been in production for four years by 1983 with conventional (IES guidelines) low rate, narrow range thermal screening had a "plug and play" reliability of only 94% (6% failed essentially out of the box). After HALT and HASS were applied using a six-axis shaker and 20°C/min air ramp rates, the plug and play jumped to 99.6% (i.e. 0.4% failed out of the box) within four months, a 15 times improvement. A subsequent power supply, which had the benefit of HALT and HASS before production began, had a plug and play of 99.7% within two months of the start of production! This company has been able simultaneously to increase sales and to reduce the QA staff from 60 to 4 mostly as a result of HALT and HASS and the impact that it had on field reliability. The company also reports that the cost of running Rel-Demo has been reduced by a *factor* of about 70 because all relevant attributable failures were found in HALT. After the application of HALT, seven products (as of 1986 when the author got the last of the data on this situation for this one company's products) had gone through Rel-Demo with *zero* attributable failures. Plug and play has been 100% since 1986!

3. In 1988, an electromechanical device was run through a series of four HALTs over a four-month period. The HALTs only took two weeks. In these tests, 39 weaknesses were found using only all-axis vibration, thermal cycling, power cycling and voltage variation. Revisions were made to the product after each HALT and then new hardware with revisions was built and then run through HALT. The designers refused to change anything unless it was verified in a life test. Extended life tests were run on 16 units for 12 weeks for 24 hours per day with three technicians present at all times to interact with the hardware. The tests revealed 40 problems, 39 of them the same as had been found in the HALTs. The HALTs had missed a lubricant degradation mode that only showed up after very extensive operation. A review of the HALT data revealed that the clues to this failure mode were in the data, but no actual failure had occurred because a technician had "helped out" and re-greased

a lead screw every night (without the author's knowledge) so that the failure that he knew about would not occur, a success in the mind of the technician at that time, before he learned what HALT was all about. His well-intended action caused an important failure mode to be missed. The author now locks up the HALT units when not actually running the tests in order to prevent well-meaning employees from "helping". Vibration to 20 GRMS all-axis random, temperatures between $-100\,°C$ and $+127\,°C$ and electrical overstress of $\pm\ 50\%$ were used along with functional testing in the HALT on the units in question. "Specs" were vibration of 1 GRMS and temperatures between 0–$40\,°C$.

4. In an informal conversation in February of 1991 between Charles Leonard of Boeing Commercial Aircraft and the author, Mr Leonard said that a quote had been received for an electronics box built under two different assumptions. The first was per the usual MIL-SPEC approach and the second was using "best practices" or the HALT and HASS approach. The second approach showed a price reduction from US$1100 to US$80, a weight reduction of 30%, a size reduction of 30% and a reliability improvement of "much better". The choice of which product to choose was obvious.

5. In 1992, the author gave a three-hour demonstration of HALT at a seminar. In this demonstration, three different products seen for the first time by the author were exposed to HALT. Each product had one failure precipitated and detected in only one hour. The products had been in field use for years with many field failures reported. All of the failures reported were exposed in the three-hour demonstration HALT. *This means that in only one hour per product all major field failure modes had been determined*. The manufacturer had not been able to duplicate the field failures using classical simulation techniques and therefore could not understand the failure modes and determine the appropriate fixes before the abbreviated HALTs were performed. Two of the three failure modes were found just beyond the edge of the temperature spec, one hot and one cold,

and the last one was found in 10 minutes at four times the "spec" GRMS using an all-axis shaker!

6. Boeing Aircraft Company reported [9] that HALT "revealed a high degree of correspondence between failures induced in the lab and documented field failures". "Vibration appears to have been the most effective failure inducement medium, particularly in combination with thermal stress". "The 777 was the first commercial airplane to receive certification for Extended Twin-engine Operations (ETOPS) at the outset of service. To a significant extent, this achievement was attributable to the extremely low initial failure trends of the avionics equipment resulting from the elevated stress testing and the corrective actions taken during development". In a conversation with Charles Leonard of Boeing in December 1995, Mr Leonard related that the "777 dispatch reliability after only two months of service was better than the next best commercial airliner after six years".

7. Nortel reported [13] a 19 times improvement in field returns when a HASSed population of PCBAs was compared with a similar population run through burn-in.

8. In 1997, a car tail light assembly was subjected to HALT costing "x". The improved assembly was run through an MTBF test costing 10 "x". The measured MTBF was 55 car lifetimes. Note that the HALT did much more good for the company than did the MTBF test and at a much lower cost. This result is fairly typical of the results of a properly run HALT, which must include corrective action. This case study is covered in [12]. The remarks above were stated later when Larry Edson was a guest speaker at one of the author's seminars.

9. In 1998, Otis Elevator reported on their web site that: "A test that would normally take up to three months to conduct can now be carried out in less than three weeks. HALT used for qualifying elevator components has saved Otis approximately US$7.5 million during the first 15 months of operation." In a conversation with the author, Otis related that a particular problem was found in a circuit board and, on one product,

corrective action was taken while on another product it was not. On the non-improved product, failure occurred after six months of use in Miami with exactly the same failure mode found in HALT. From this the author jokingly says that two days in HALT is equivalent to six months in Miami!

1.9 SOME GENERAL COMMENTS ON HALT AND HASS

The successful use of HALT or HASS requires several actions to be completed. In sequence these are: precipitation, detection, failure analysis, corrective action, verification of corrective action and then entry into a database. *All* of the first five must be done in order for the method to function at all. Adding the sixth results in long-term improvement of the future products.

1. *Precipitation* means to change a defect which is latent or un-detectable to one that is patent or detectable. A poor solder joint is such an example. When latent, it is probably not detectable electrically unless it is extremely poor. The process of precipitation will transpose the flaw to one that is detectable; that is, cracked. This cracked joint may be detectable under certain conditions, such as modulated excitation. The stresses used for the transformation may be vibration combined with thermal cycling and perhaps electrical overstress. Precipitation is usually accomplished in HALT or in a precipitation screen.

2. *Detection* means to determine that a fault exists. After precipitation by whatever means, it may become patent that is, detectable. Just because it is patent does not mean that it will actually be detected since it must first be put into a detectable state, perhaps using modulated excitation, and then it must actually be detected. Assuming that we actually put the fault into a detectable state and that the built-in test or external test setup can detect the fault, we can then proceed to the most difficult step, which is failure analysis.

3. *Failure analysis* means to determine why the failure occurred. In the case of the solder joint, we need to determine why the joint

failed. If doing HALT, the failed joint could be due to a design flaw; that is, an extreme stress at the joint due to vibration or maybe due to a poor match of thermal expansion coefficients. When doing HASS, the design is assumed to be satisfactory (which may not be true if changes have occurred) and, in that case, the solder joint was probably defective. In what manner it was defective and why it was defective need to be determined in sufficient detail to perform the next step, which is corrective action.

4. *Corrective action* means to change the design or processes as appropriate so that the failure will not occur again in the future. This step is absolutely essential if success is to be accomplished. In fact, corrective action is the main purpose of performing HALT or HASS.

5. *Verification of corrective action* needs to be accomplished by testing to determine that the product is really fixed and that the flaw which caused the problem is no longer present. The fix could be ineffective or there could be other problems causing the anomaly which are not yet fixed. Additionally, another fault could be induced by operations on the product and this necessitates a repeat of the conditions that prompted the fault to be evident. Note that a test under zero stress conditions will usually not expose the fault. One method of testing a fix during the HALT stage is to perform HALT again and determine that the product is at least as robust as it was before and it should be somewhat better. If one is in the HASS stage, then performing HASS again on the product is in order. If the flaw is correctly fixed, then the same failure should not occur again.

It is essential to have at least the first five steps completed in order to be successful in improving the reliability of a product. If any one of the first five steps is not completed correctly, then no improvement will occur and the general trend in reliability will be toward a continuously lower level.

6. The last step is to put the lesson learned into a database from which one can extract valuable knowledge whenever a similar

Table 1.1 Comparison of HALT and Classical Approaches

Stage	Design		Pre-production			Production			
Test type	Quality	HALT	Life test	HASS development	Safety of HASS	Rel-Demo	HASS	HASS Optimization
Purpose	Satisfy Customer reqmts	Maximize margins, minimize sample	Demo life	Select screens and equip	Prove OK to ship	Measure reliability	Improve reliability	Minimize cost, maximize effectiveness
Desired outcome	Customer acceptance	Improve margins	MTBF and spares reqd	Minimize cost, maximize reliability	Life left after HASS	Pass	Root cause corrective action	Minimize cost, maximize effectiveness
Method	Simulate field environment sequentially	Step stress to failure	Simulate Field	Maximize time compression	Multiple repeats without wearout	Simulate field	Accelerated stimulation	Repeat HASS, modify profiles
Duration	Weeks	Days	Weeks	Days	Days	Months	Minutes	Weeks
Stress level	Field	Exceeds field	Field	Exceeds field	Exceeds field	Field	Exceeds field	Exceeds field

event occurs again. Companies which practice correct HALT and utilize a well-kept database soon become very adept at designing and building very robust products with the commensurate high reliability. These companies usually are also very accomplished at HASS and so can progress to HASA, the audit version of HASS.

A comparison of the HALT and HASS approach and the classical approach is presented in Table 1.1. Note that HALT and HASS are proactive, i.e. seek to improve the product's reliability, and much of the classical approachs are intended to measure the product's reliability, not to improve it.

1.10 SUMMARY

Today, HALT and HASS are required on an ever-increasing number of commercial and military programs. Many of the leading commercial companies are successfully using HALT and HASS techniques with all-axis broad band vibration and moderate to very high-rate thermal systems. However, most are restricting publication of results because of the phenomenal improvements in quality and reliability and vast cost savings attained by using the methods. The aerospace and military have been slow to accept the advanced techniques since stressing the product above the expected field environment is quite foreign to them, and even if they wanted to use HALT and HASS, it is very difficult to specify or to interpret in contractual language. In recent years, there has been a turn toward HALT and HASS on US military programs. They have discontinued the use of MIL-SPECS for environments and added very long warranty periods, 15–20 years being typical now.

One has to want to obtain top quality in order to adopt the cultural change necessary for the adoption of HALT and HASS. The basic philosophy is, simply stated, "find the weaknesses however one can and use these discoveries as opportunities for improvement". This constitutes a new paradigm!

Some contracts, both military and civilian, erroneously require the use of MIL-HDBK 217 or similar reliability prediction methods and, therefore, restrict the program to use incorrect and very misleading predictors of MTBF. See [7] for very good discussions of the reasons behind this statement. HALT and HASS focus on *improving* reliability, not on measuring it.

The best known method to determine the actual reliability of a product is to test numerous samples of the product in field environments for extended periods of time. This would, of course, either delay the introduction of the product or provide reliability answers far too late in order to take timely corrective action. Significant time compression can be obtained by eliminating low stress events which create little fatigue damage and simulating only the high stress events. This approach may reduce the test time from years to months. The HALT and HASS approach accelerates this even further by increasing the stresses to far beyond the actual field levels and decreases the time to failure to a few days or even hours, sometimes only a few seconds.

Later chapters will go into considerably more detail about the various aspects of the HALT and HASS technologies.

Some product lines to which HALT and HASS have been successfully applied are listed in Table 1.2. This list is not all-inclusive due to many non-disclosure agreements.

REFERENCES AND NOTES

1. O'Connor, P. D. T., *Practical Reliability Engineering*, 2nd Ed., John Wiley & Sons (1985).
2. Bralla, J. G., *Design for Excellence*, McGraw-Hill (1996), p. 255.
3. Seusy, C., Achieving phenomenal reliability growth, *ASM Conference on Reliability – Key to Industrial Success*, Los Angeles, CA, 24–26 March 1987.
4. Hobbs, G. K., Development of stress screens, *Proceedings of the 1987 Reliability and Maintainability Symposium*, p. 115. This paradigm-shifting paper was rejected by the IES for presentation at the 1986 Technical Meeting and therefore was presented at RAMS the next year.
5. Institute of Environmental Sciences, *Environmental Stress Screening Guidelines* (1981). This document is very misleading as "effectiveness" is confused with tradition, contracts and popularity.

28 INTRODUCTION TO HALT AND HASS

Table 1.2 Some products successfully subjected to HALT and HASS

ABS systems	Centrifuges	Locomotive engine controls
Accelerometers	Check canceling machines	Locomotive electronics
Air conditioners	Circuit boards	Loran systems
Air conditioner control systems	Climate control systems	Magnetic resonance instruments
Air bag control modules	Clothes washing machines	Manual transmissions
Aircraft avionics	Clothes dryers	Mainframe computers
Aircraft flap controllers	Computers	Medical electronics
Aircraft hydraulic controls	Computer keyboards	Meters
Aircraft instruments	Communication radios	Microwave communication systems
Aircraft pneumatic controls	Copiers	Microwave ranges
Aircraft engine controls	Dialysis systems	Missiles
Aircraft antenna systems	Dish washers	Modems
Anesthesiology delivery devices	Disk drives	Monitors
Anti skid braking systems	Distance measuring equipment	MRI equipment
Area navigation systems	Down hole electronics	Navigation systems
Arrays of disk drives	Electronic controls	Notebook somputers
ASICs	Electronic carburetors	Oscilloscopes
Audio systems	Electronics	Ovens
Automation systems	FAX machines	Oximeters
Automotive dashboards	Fire sensor systems	Pacemakers
Automotive engine controls	Flight control systems	Personal computers
Automotive exhaust systems	Flow sensing instruments	Plotters
Automotive interior electronics	FM tuners	Pneumatic vibration systems
Automotive speed controls	Garage door openers	Point-of-sale data systems
Automotive traction controls	Global positioning systems	Portable communications
Blood analysis equipment	Guidance and control systems	Portable welding systems
Calculators	Heart monitoring systems	Power tools
Cameras	Impact printers	Power supplies
Card cages	Ink jet printers	Power control modules
Casagranian telescope structure	Instant cameras	Printers
Cash registers	Invasive monitoring devices	Prostate treatment system
Cassette players	IV drip monitors	Proximity fuses
CAT scanner	Jet engine controllers	Racks of electronics
CB radios	Laptop computers	Radar systems
	Laser printers	Refrigerators
	Lipstick	Respiratory gas monitors
	LN2 thermal cycling chambers	Safety and arming devices

Table 1.2 *Continued*

Shaker tables	Televisions	Turbine engine control modules
Solid state memory systems	Thermal control systems	Typewriters
Spectrum analyzers	Thermal imaging gun sight	Ultrasound equipment
Speed brake controls		Urine analysis machines
Stationary welding systems	Thermostats	Vibration control systems
Stereo receivers	Torpedo electronics	
Switching power supplies	Traction control systems	Vibration monitoring systems
Tape drive systems	Tractor engine control modules	
Tape players		Vibrators
Target tracking systems	Tractor instrumentation	Video recorders
Telecommunications equipment	Transmission controls	Vital signs monitors
	Trash compactors	Water sprinkler systems
	Turbine engine monitoring equipment	Work stations
Telephone systems		X-ray systems

Note: A list of current users who are willing to discuss their results is available from Hobbs Engineering Corporation, 4300 West 100th Avenue, Westminster, CO 80031-2481, (303) 465-5988, Fax (303) 469-2130, e-mail: learn@hobbsengr.com, Web: hobbsengr.com. This list changes often and so cannot be published and remain in current form.

6. Hobbs, G. K., Screening technology and accelerated methods, *Proceedings of the IES/ESSEH Workshops on Accelerated Stress Applications, Vancouver*, WA, 17–19 March, 1992.
7. *Quality and Reliability Engineering International*, **6**, (4), Sept–Oct (1990), John Wiley & Sons. There are many excellent articles on failure prediction methodology in this issue. This issue is must reading for anyone using MIL-HDBK-217 type methods.
8. Hakim, E. B., Microelectronic reliability/temperature independence, US Army LABCOM, *Quality and Reliability Engineering International*, **7**, pp. 215–220, (1991).
9. Miner, M. A., Cumulative damage in fatigue, *Journal of Applied Mechanics*, **12**, 1945.
10. Hobbs, G. K. and Holmes, J., Tri-Axial vibration screening – an effective tool, *IES/ESSEH*, San Jose, CA, 21–25 September, 1981.
11. Smithson, S. A., Effectiveness and economics – yardsticks for ESS decisions, *1990 Proceedings of the IES*.
12. Edson, L., Combining team spirit and statistical tools with the HALT process", *Proceedings of the 1996 Accelerated Reliability Technology Symposium*, Hobbs Engineering Corporation, Denver, CO, 16–20 September 1996.
13. Department of the Navy, NAVMAT P-9492, May 1979.

14. Bernard, A., The French environmental stress screening program, *Proceedings of the 1985 IES, 31st Annual Technical Meeting*, pp 439–442. This document is very misleading because "effectiveness" is confused with tradition, contracts and popularity.
15. Cooper M. R. and Stone, K. P., Manufacturing stress screening results for a switched mode power supply, *1996 Proceedings of the IES*.
16. Hobbs, G. K., Stress screening today, presented at the *Quality Expo Time West*, Los Angeles, CA, 11 November, 1985.
17. Hobbs, G. K., Recent advances in stress screening, *40th Annual Quality Congress*, Anaheim, CA, 19–21 May, 1986.
18. Hobbs, G. K. "Stress screening: Progress setbacks and the future, *REL-CON Europe '86*, Copenhagen, Denmark, 16–20 June, 1986.
19. Hobbs, G. K., Deceptive stress screening and how to detect it, *Sound and Vibration*, September 1987.
20. Hobbs, G. K., Why the Leaders in Screening Are Not Publishing, Hobbs Engineering Corporation (1988).
21. Minor, E. O., Accelerated quality maturity for avionics, *Proceedings of the 1996 Accelerated Reliability Technology Symposium*, Hobbs Engineering Corporation, Denver, CO. 16–20 September, 1996.
22. O'Connor, P. D. T., *Engineering Management, A New Approach*, John Wiley & Sons (1994).
23. Wong, K., A new environmental stress screening theory for electronics, *1989 Proceedings of the IES*, pp. 218–224.

CHAPTER 2
HALT: Highly Accelerated Life Tests

*Every truth passes through three stages before it is recognized.
In the first, it is ridiculed,
In the second, it is opposed,
in the third, it is recognized as self evident.*

Arther Schopenhauer,
Nineteenth-century German Philosopher

2.1 INTRODUCTION

A test in which stresses applied to the product are well beyond normal shipping, storage and in-use levels is called a *Highly Accelerated Life Test*, for which the acronym HALT has been selected. HALTs are run for several reasons and are an extremely valuable part of forced design and process maturation. Design as used herein is meant to include all of the testing included in the design phases of a product's development. HALTs are also required precursors to Highly Accelerated Stress Screen (HASS) development. HALT will generally result in reduced design development time and cost, improved field reliability and all of the benefits that improved field reliability brings to fruition, such as improved customer satisfaction, increased market share, economy of scale and consequent improved profitability. Several points should be made regarding design margins, sample size required to measure the design margins and process control.

1. Design margins are impossible to measure accurately without testing many units. The basis for this statement is that any one variable property of a device will be a distribution over the population, and the exact shape of this distribution will be unknown ahead of time. The statistical properties of any distribution must be measured by testing a large sample of the population in order to obtain a statistically significant estimate of the distribution. The *tails* of the distribution are the primary interest in HALT and HASS. This was shown in graphical form in Figure 1.1 in Chapter 1. It was the low-strength tail that caused the field failures. An alternate method to the large sample would be to use accelerated environments and other accelerated stimuli, such as line voltage or frequency on a small sample in order to be confident of reasonable design margins for the entire population. Using the step stress techniques, the stress levels will be substituted for sample size.

When the stresses are increased, the distributions are compressed [1]. This is due to the fundamental physics of failure and may be known to the reader from studying the strength of materials. In tensile testing, the higher the stress, the more narrow the distribution over the population. It is essential to realize that when the stress levels are increased, time compression occurs and that the failure modes usually do not change unless a change of state occurs. Increasing the temperature up to the glass transition temperature of a printed circuit board would be such an event. In that case, the failure modes definitely would change from those found below the glass transition temperature.

The failure modes exposed by HALT do not necessarily occur in the same order that the field Pareto chart would show, but, generally, all of the significant modes of field failures are exposed in a properly done HALT. It is also important to note that some failure modes will be found with a different stress than the one that would cause the failure in the field. Figure 1.6 in Chapter 1 illustrates this important fact of life. In that figure, the horizontal axes have been adjusted so that both of the curves align in terms of field stress and the stress in HALT where a particular failure is observed. It is seen that, for the particular failure mode illustrated, the highest failure rate under field conditions will be due to temperature, whereas, in HALT, the highest failure rate will be due to vibration. This means that with a small sample as in HALT, the failure mode will likely be precipitated in vibration and, in the field, most likely will be precipitated in temperature. This example is just an illustration and is not based on any real flaw. Concentration on the stress level used would lead one to conclude that an opportunity for improvement was not present, only an opportunity to spend more money on a change that would not improve field performance. This mistake is frequently made by engineers not properly trained in the underlying principles of HALT.

2. Early prototype hardware will generally include both design and process problems. A very large sample would be required in order to gain statistical significance in the measurement of the flaw types present. If only a few units in a group have a given problem, perhaps hundreds of units would have to be tested in order to spot these few problems. Cliff Seusy's paper [2] addresses both the design and process questions and gives some estimates of the number of units required to be tested under typical conditions in order to find flaws with an estimated confidence level with a population of a given percent defective if typical stress levels are used. For example, if one observation of a given problem were enough to convince the investigator of its relevance, if 2% of the population had a given problem, and if a 95% probability of detection were desired, then Seusy's paper shows that about 145 units would be needed for test (for *each* failure mode)! Clearly, accelerated techniques with the concomitant reduction in sample size is essential for a cost-effective reduction of design flaws.

3. The capability to detect flawed performance by built-in self-test or by external test systems is called coverage. Coverage is usually small; that is, much less than 100% during the early stages of design and test. This low coverage will severely limit the ability to find, and consequently improve, any design deficiencies. Resolution is the capability to pinpoint the area of the fault. Both high coverage and good resolution are essential for the improvement of any design weak links since, without

them, few flaws will be detected and located, respectively; therefore, few can be eliminated. Chapter 8 describes this subject in some detail and introduces Software HALT, a method for coverage and resolution improvement that is performed separately from the subject of the rest of this book. That is, Software HALT is performed as a software development and improvement activity.

4. HASS requires that the product has a generous margin of strength for the short-term application of stresses above those which occur in a typical use of the product. These margins are attained by applying the HALT approach. The stresses may include temperatures, vibration levels, voltages and other stimuli which exceed the normal levels. These very high levels are used in HASS to force rapid defect precipitation in order to make the screens rapid, effective and economical. The use of HASS requires knowledge about the product's ability to function at extended ranges of stimulation and knowledge about the failure mechanisms which limit the stimuli levels. Design and process changes are usually made to extend the functional and destruct levels of the equipment in order to assure large design and process margins which lead directly to reduced field failures as well as to allowing HASS with its attendant cost savings and reliability improvements. Safety of HASS will be used to demonstrate that repeated screening does not result in product end of life; that is, that a few passes through HASS only remove a very small percentage of the total life in the product, leaving more than enough for a normal field lifetime.

2.2 DEFINITION OF TERMS

The *upper operational limit* is defined as that stress level or dimension or other variable above which the product will not operate properly. "Properly" may mean within specifications or may have another definition as appropriate to the situation.

The *lower operational limit* is defined as that stress level or dimension or other variable below which the product will not operate properly.

The *upper (lower) destruct limit* is that stress above (or below) which the product will experience a change that precludes operation even when the stress is reduced to less than (more than) the operational limit. The destruct limit is that stress level above or below which the product will suffer permanent failure that does not recover when the stress is reduced.

Both of these will be discussed more thoroughly later.

2.3 STIMULI APPLIED IN HALT

The tests begin with a time-compressed investigation of the product's design and process limits of operation and destruct. Step stress testing in generic stresses, such as vibration, temperature, rate of change of temperature, voltage, power cycling and humidity is done. These stresses are generally useful for most products. In addition, product unique stresses, such as clock frequency, d.c. voltage variation and even component value variation may be used as appropriate for the product under test. Other variables which may be used include contaminants in fluids or air, the pH of a fluid, the viscosity of a fluid, the outside diameter of a gear and any other stress or variable that could expose design or process limitations or sensitivities.

For all of these stimuli, the upper and lower operational and destruct limits should be found and/or understood. By "understood" is meant that although the limits are not actually found, they are verified to be well beyond the limits used in HASS and even farther beyond the field environments. For example, a product may be able to withstand an hour of simultaneous all-axis random vibration at 30 GRMS without failure. Although the destruct limit has not been found, it is certainly high enough for most commercial equipment intended for non-military environments and where the screen environment may be 10 GRMS all-axis random for five minutes and the worst field environment a truck ride while in a container that provides vibration isolation. This example of capability is quite common when HALT is properly applied. One manufacturer of computer peripherals regularly runs production screens at vibration levels which would be above destruct for most military avionics! The rugged products supplied by this company have attained a premier position in the world as the most reliable of any in their field. Large margins translate directly into high reliability; however, knowing the margins does not directly lead to a calculation of field reliability. One would have to know the distributions of all operational and destruct modes in order to calculate field failure rates. This would relate to design factors only, not to quality problems.

There are several reasons for ascertaining the operational and destruct limits. Knowledge of the operational limits is necessary in order to assess whether design margins exist and how large they are likely to be on a population basis. It is also necessary to determine whether detection tests can be run during HASS since the detection tests run *during* stimulation are necessary for high detectability of precipitated defects. In addition, test coverage and resolution must be acceptably high in order to find the flaws that are precipitated. See Chapter 8 for more details on this subject. Knowledge of the destruct limits is required in order to determine design margins on the non-operating environments and to select HASS environments below known destruct levels. See Chapter 3 on HASS for the various types of screens and when to apply them. *Note that precipitation and detection screens are used during HALT, therefore Chapter 3 should be mastered before attempting HALT.* If this is not done, then many opportunities for improvement will be missed.

Some stimuli and the associated operational and destruct levels are briefly discussed below. There is no significance to the order of discussion and alphabetical order is used for the first three stimuli. Generally speaking, all-axis broad-band vibration is the most effective in terms of the number of design and process flaws exposed in electronic equipment if only one stress is used; however, the simultaneous application of many stresses is even more effective. This is discussed at length in the next chapter. Recall the discussion in Chapter 1 where it was mentioned that temperature had been rated as the most effective screen based on a survey of contractors who were mostly directed to do thermal cycling by contract and then asked to rate screens, but only rate those which had been used. This seems to be circular logic.

In the discussions that follow, temperature is discussed first because it lends itself to the discussion and many products have all of the limits discussed. If vibration were to be used first, then the concept of lower operational and destruct limits could not be introduced as one cannot vibrate at negative GRMS values. It is reiterated that temperature is usually not the most effective stress in the finding of weaknesses on many products, including electronics [3].

It is very important to note that modulated excitation, discussed in the next chapter, is frequently required in order to detect a patent defect. That is, during HALT activities, one needs to perform modulated excitation in order to detect the fact that a failure has occurred. It is, therefore, necessary to read and understand the next chapter in order to perform HALT properly. Briefly stated, Modulated Excitation consists of a slow variation of the temperature while performing simultaneous rapid variation of the overall six-axis vibration level. This constitutes a search pattern in the two-space of temperature and vibration. It frequently happens that a patent defect is only detectable at some small area in the two-space. That is why modulated excitation is used for detection during HALT.

2.3.1 Temperature

The Upper Operational Temperature Limit (UOTL) is that temperature below which the unit will operate completely normally; that is, within all normal operating parameters. This temperature should be well above the expected typical operational temperature environment in order to ensure that the whole population will work as intended at the high operational temperature. This temperature must also be known in order to select a detection screen, since above this temperature the unit will not run normally and this will not be considered to be a failure in HASS. Discriminators can possibly be used above the UOTL to sort the defect-free products from the flawed products even though the system may not operate correctly at temperatures above the UOTL. A discriminator is a test or method to determine which units are defective and which are flaw-free although, during the discrimination, none of the units may operate within spec.

The Upper Destruct Temperature Limit (UDTL) is that temperature above which the unit will fail permanently. This temperature should be well above the expected maximum field temperatures including storage and transportation in order to ensure that the whole population will survive the field environment. This information is also required in order to set the maximum precipitation screen level and to ensure that the screen is not destructive to good hardware. The latter is also proved by the technique of Safety of HASS, covered in Chapter 3.

The Lower Operational Temperature Limit (LOTL) is the lowest temperature above which the unit will operate completely normally. This temperature should be well below the expected minimum operational temperature environment in order to ensure that the whole population will work as intended at the low operational temperature. This temperature must also be known in order to select a detection screen, since below this temperature the unit will not operate normally. Discriminators can possibly be used below the LOTL to sort the defect-free products from the flawed ones.

The Lower Destruct Temperature Limit (LDTL) is that temperature below which the unit will fail permanently. This temperature should be well below the expected minimum field temperatures including storage and transportation in order to ensure that the whole population will survive the lowest temperatures. This information is also required in order to set the minimum precipitation screen temperature level and to ensure that the screen is not destructive to defect-free hardware.

A typical way in which to proceed might be to check the product's operation within the "spec" limits of temperature. These may be +10°C to +40°C if the product were a computer peripheral for office use. For a military product, the "spec" environment may be −55°C to +80°C. At any rate, the product has some "spec". The quotation marks are used here since the real in-use environment is often not known with any certainty, nor is it of great importance or interest when using the HALT method.

The usual procedure is to raise the temperature perhaps 10°C at a time, *while continuously monitoring*, until the product fails to operate correctly. Fine resolution of the temperature at which the product fails to function is not needed since that level will be improved and, in addition, only one product in the population is being stressed at this time and there will be variation over the population. A long dwell is not necessary since the product's behavior will most likely be modified when the feature limiting the operational extension is determined. Another reason for not dwelling is that the product temperature is controlled, not chamber temperature, using modern control systems, so a stabilization dwell is not necessary unless some time-at-temperature related failure mode such as solder joint creep is being investigated. Many other failure modes of interest are not time related with the exception of those following the Arrhenius equation, such as chemical reactions and electromigration [4].

In the next step of the thermal HALT, the product temperature is reduced a few degrees at a time until the correct operation returns, if it does return at all. If it does, then an operational limit has been found. If it does not, then a destruct limit has been found. For the sake of discussion, assume that an operational limit has been found. Then, an attempt to modify the product is made so that it will operate at a higher stress level, in this case, an upper operational temperature. The difference between the "spec" and the upper operational limit defines the upper operational margin. It is known that large margins relate directly to low field failure rates in some unknown, but directly correlated, fashion. Therefore, it is beneficial to increase the margin as much as possible, right up to the "fundamental limit of the technology". The fundamental limit of the technology means that it is not known how to make it any better *within reasonable technical and cost constraints*. A corollary limit is that imposed by the business situation of being out of time or money. In that case, production may begin with known (or even unknown) design margin limiters or robustness limiters which may be thought of as design flaws or opportunities for improvement which remain undiscovered or unexercised. It is suggested that one should not be too quick to declare that a fundamental limit has been reached or that plenty of margin exists, since it is very expensive to do so in error. The average cost observed by the author for doing so incorrectly at Hewlett-Packard [3] and other large-volume producers is in the neighborhood of US$10,000,000/ flaw not corrected! A plethora of features can usually be improved for a fraction of this amount! It therefore makes sense to take advantage of all opportunities for improvement as long as the expense is minimal.

It is best not to focus on the stress used to expose the design flaw, but to focus on the failure mode and mechanism instead. Figure 1.6 in Chapter 1 shows the instantaneous failure rate versus stress level for some hypothetical product. The curves for stress 1 and stress 2 have been adjusted so that the field environment and the HALT environment at which a weakness is found of each one is at the same abscissa. What is shown is that a failure mode (opportunity for improvement) that is exposed first in HALT by one stress may not be exposed first in the field by the same stress. This is why the whole battery of possible stresses including combinations of all stresses should always be used. Concentration should be on the failure modes and mechanisms instead of asking: "Where will the product ever see this stress level?" Thinking about margins, too, will lead one astray as one can readily see in the above example. It may be that the margin is very large considering the one stress being applied; however, some other stress may have a much lower margin for the different stress which could result in field failures. It should be assumed that all failures are relevant and treated as opportunities for improvement. A number of products have been exposed to HALT by the author wherein a weakness has been found by some stress other than the one causing the problems in the field. If you find it, it is probably relevant regardless of how you found it. HALT only points a big red arrow at the weakness.

The general rule of thumb in correctly applied HALT is to just keep improving the product and making it more robust until the cost per fix becomes inordinately high for your product or *all* the margins, including those for combined stresses, are obviously excessive. A point to be made is that all of the products in the population will not stop operation at the same temperature (or other stress) and that a distribution of operational limits over the population for any given failure mode will exist. Many failure modes may affect the operation and each should be investigated separately. The shape of the distributions will not be known in the HALT stage of operation since one will be working with engineering prototypes and will have only a few of them. One way to mitigate ignorance of the distribution shape is to just make the margins very large, somewhat alleviating the question of distributions entirely.

Several situations may result in different distribution shapes. There can be broadly distributed failures such as in Figure 2.1, narrowly distributed failures with large margins as in Figure 2.2, narrowly distributed failures with small margins as in Figure 2.3 and multi-modal distributions as in Figure 2.4. Note that, in general, one will find many opportunities for improvement with various distribution patterns, whereas the discussion here has just

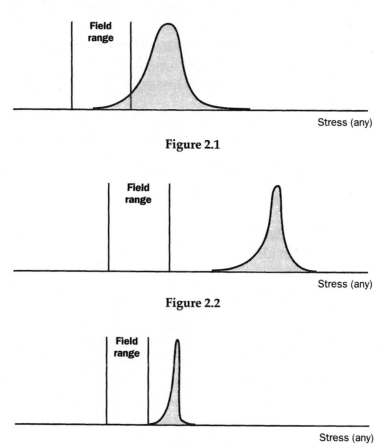

Figure 2.1

Figure 2.2

Figure 2.3

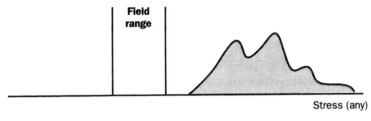

Figure 2.4

touched upon one failure mode. There are, in general, many failure modes of any product and one should investigate all of them for potential improvement opportunities. Some distributions are very wide and may not be observed unless very high stresses are used. In HALT, very high stresses are used and, in effect, are a substitute for a large sample size. As long as one can observe one failure out of a distribution, then that distribution can be moved to a higher stress level; that is, an opportunity for improvement is exercised. It is imperative that all failures be investigated without consideration of the margin.

The "Six Sigma" approach is one wherein the manufacturers attempt to control processes very tightly so that variations of six standard deviations from the mean will satisfy the design requirements. The narrow distributions with small margins as in Figure 2.3 may result from the Six Sigma approach which has been implemented without HALT. The situation shown in Figure 2.3 would be unacceptable because, if a process were to go out of control by just a small amount, field failures would occur. In the case of a distribution of the shape of Figure 2.2 with large margins, a process change would not necessarily cause field failures since large margins would exist and the product would be tolerant of parameter changes to a large degree. This points out a fundamental difference between the HALT and the Six Sigma approaches: HALT achieves large margins and robust products which are flaw tolerant. The Six Sigma approach used prior to or without HALT narrows the distributions before or without attaining large margins, respectively. Using HALT to achieve large margins and *then* using the Six Sigma approach to process control is the optimum way to achieve reliability. Large margins are the primary goal of HALT with the secondary goal being to obtain the data necessary in order to begin HASS. The attainment of large margins will result in very substantial gains in field reliability although no measure of that reliability is obtained. However, the product will be designed as well as possible.

After identifying a design or process flaw (opportunity for improvement), we remove the flaw by a design or a process change, thereby obtaining larger capability. During HALT, this may be done on a temporary basis. For example, if some screws back out, Loctite is applied. If some large component flies off the circuit board, then the component is replaced and bonded to the board with quick hardening epoxy, and so on. Later, permanent design or process changes will be performed. The strength distribution for that flaw will therefore be moved to the right. The upper operational limit is then explored again and another design or process limiter could be discovered. The new distribution is moved to the right by some improvement in the design or process and so on until the point is reached where the fundamental limit of the technology has been reached. By then, one could have moved many distributions to the right and exercised many opportunities for improvement.

After pushing the upper limit of the operational envelope, one now starts to explore for the lower operational limit using the step stress technique. The temperature (or other stress) is decreased until operation ceases. Then an increase in temperature a few degrees at a time is performed until operation returns to normal and defines the lower operational limit. If operation does not return, then a lower destruct limit has been discovered. Again, there is a distribution over the population and one will attempt to limit risk by making the lower operational margin as large as possible within the realm of good engineering practice (which is not just meeting "spec"). Each time an operational limit is ascertained, we attempt to push the product to the fundamental limit of the technology and move the distribution to the left.

The next step in the thermal HALT is to look for and improve the destruct limits at high and low temperatures. The usual approach is to look for the lower limit (in temperature) first as many products do not have a destruct limit above $-100\,°C$ in the author's experience. If the product's destruct limit is below −100 °C, there is usually little interest in extending it since most field environments are well above this low limit and low temperatures accelerate few failure modes except those due to mismatched thermal expansion coefficients which usually takes a few broad-range cycles to precipitate. Performing the lower limit investigation first usually does not result in any repair required and allows the accumulation of as much information about the product as possible before failures start to occur and repairs of a temporary nature have to be made. At some point, a failure that cannot be easily repaired on the test unit will occur and another unit to test might be needed. It is desirable to obtain as much product knowledge as possible before an irreparable failure is caused since usually only a few (or maybe only one) samples are available.

Finally, the upper destruct limit is investigated by stepping up in temperature (or other stress) until return to within the operational region after exposure to the high temperature will not return the normal function of the product. The destruct limits also have a distribution over the population as do the operational limits. Unawareness of the distribution is overcome by maximizing the margin. Assurance that the product is robust and therefore flaw tolerant is attained by doing so. This can be considered to be severe derating. Figure 2.5 illustrates the margins obtained.

In addition to using temperature in HALT, many other stresses are used as well, including, but not limited to, vibration, voltage, current, temperature, temperature rate of change, humidity and

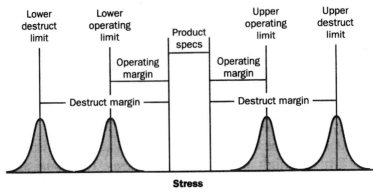

Figure 2.5

any other stress which could show up relevant design or process problems in the product under test. *Note that no one stress can find all of the design and process problems in most products!* Experts in HALT use their entire repertoire singly and in appropriate combinations to find all design and process robustness limiters. The IES guidelines already mentioned in Chapter 1 did severe damage to the progress of ESS in the United States by calling thermal cycling the "most effective" screen. Data discussed later in this chapter [3], and indeed much other data, shows that all-axis vibration is far superior to thermal cycling in uncovering design and process flaws in most equipment, including electronics. This does not mean that all-axis vibration alone should be used. The point is made that all appropriate stresses should be used, not just one.

Failures precipitated in temperature (or any other stress) may not be detectable unless modulated excitation is performed. At this stage, before any vibration-related limits have been established, one can just use low-level vibration in the modulated excitation.

Rate of Change of Temperature

The rate of change of temperature step stress testing was discontinued by the author after taking several days to perform a thermal ramp rate step stress test back in 1983. Nothing was found to be rate sensitive in that series of tests and nothing has been found by the author since. This statement is made based on the experience of running HALTs on over 100 products. However, this limited experience does not ensure that no rate-sensitive limiters exist in any specific product since indeed, several have been found and reported to the author by attendees at his seminars. It is therefore suggested that the maximum thermal ramp rate is run as a first test, and then, if anything fails, a step stress test should be run to determine the sensitivity of the product to ramp rate. Any ramp rate limiter should be eliminated if at all possible since the total cost of production screening tends to be inversely proportional to the ramp rate used to a substantial degree. It may be far less costly to modify the product to withstand higher ramp rates than to increase the number of personnel, thermal chambers, shakers and test setups to accommodate the lower ramp rates with the concomitant costs. Also, some real in-use ramp rates on autos and airplanes are very high, up to 100°C/min, and the associated products must be able to withstand this high ramp rate for many cycles if reliable operation is to be attained. For example, the under-hood ramp rate on some cars reaches –60°C/min when an automobile sits at a stop light in the summer and then accelerates down an on ramp onto the freeway. Some ram air-cooled aircraft avionics experience ± 100°C/min and recurring multiple times per day as well.

Again, note that modulated excitation may be necessary to detect patent defects.

2.3.2 Vibration

The Operational Vibration Level (OVL) is the highest vibration level at which the unit will operate completely correctly. This level should be well above the expected normal operational vibration environment in order to ensure that the *entire* population will all work as intended at the highest OVL. This level must also be known in order to select a *detection screen,* since above this vibration level the product will not operate correctly. Discriminators can possibly be used above the OVL to sort the defect-free ones from the flawed ones. A microphonic signature analysis has been found to be an effective discriminator in several cases.

The Vibration Destruct Level (VDL) is that vibration level above which the unit will fail permanently. This level should be well above the expected worst case field level including transportation in order to ensure that the entire population will survive. This information is also required in order to set the maximum precipitation screen level and to ensure that the screen is not destructive to good hardware.

In determining the OVL and VDL, it is important to use the same type of shaker for the investigation as will be used in the screens since the fatigue damage accumulation rate is very much higher for the impact type of shakers than it is for the electrodynamic or servo hydraulic types used at the same GRMS level. See Chapter 10 for more details. This difference is viewed by the author as a plus for the impact type of shakers since this property allows much faster screens and therefore requires substantially less equipment, including monitoring test equipment. The spectrum shapes, probability densities and x–y–z balances are also different on the various brands and types of shakers. Note that the monitoring test equipment frequently costs much more than the shakers and chambers.

The fixturing should provide the same boundary conditions (method of mounting) to the product as will exist in field usage so that the natural frequencies and mode shapes are the same as in a service environment. This will provide high stresses where the field environment provides high stresses and low stresses where the field environment provides low stresses. That is, the stressing will expose what the field environment would expose. Any other boundary conditions will not provide good replication of field failures and so should not be used. "Boundary conditions" as used here means constraints as in edge constraints either clamped, restrained from displacement only or completely free.

The procedure is the same as discussed before in temperature. The vibration is stepped up in increments until either an operational or a destruct limit is found. A means of increasing the limit is then introduced in order to move that distribution to the right and then the investigation continues up until the fundamental limit is reached.

Modulated excitation may be necessary to detect patent defects.

2.3.3 Voltage

For electrical and electronic units, the Upper and Lower Voltage Operational Limits (UVOL and LVOL, respectively) are the upper and lower voltages at which the unit will operate properly. These characteristics are required to determine design margins and to determine the parameters in a detection screen. These operational limits are usually also affected by the frequency of the supply voltage and by the temperature as well.

48 HALT: HIGHLY ACCELERATED LIFE TESTS

The Upper Voltage Destruct Limit (UVDL) is the maximum voltage which can be applied for a given period of time without unit failure. This property may be affected by the temperature and perhaps also by the frequency of the voltage if it is not DC. The UVDL is needed to assess design margins and to select a non-destructive screen level in HASS. The Lower Voltage Destruct Limit (LVDL) may or may not exist depending on the product.

Voltage testing may also contain stresses such as electrostatic discharge and lightning strikes. Every stress to which the product may be sensitive is used in order to find the weak links in the product.

2.3.4 Four Corner Tests

Consider Figure 2.6 which portrays a four corner test combing any two stresses. The old technique was to test only at the corners and then imply the product would work properly anywhere within the "box". This has been frequently found not to be true. Illustrated is a case wherein there are two areas within which the product will not function properly. In the next Chapter, modulated excitation, which has the purpose of searching the entire box for inoperative areas, will be discussed. The author has applied the search method

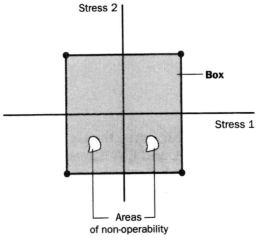

Figure 2.6

to temperature and vibration with outstanding success. In principle at least, this method can be applied to other stresses such as voltage and frequency and other combinations as well.

2.3.5 Other Stresses

Many other product unique stresses or parameters may be defined as appropriate for a particular product. These may include humidity, the pH of a solution in a product, the salinity or the viscosity of a fluid running through the equipment, the amount or size of particulate matter in a fluid, gear condition, lubricant, shaft misallignment, unbalance and many other stresses or conditions. Any test or stress that reveals design or process problems is appropriate. Note in particular that the stimuli need not be representative of expected field environments, they need only to expose the types of failures that would occur in the field. HALT is most definitely not a simulation, but a search for any kind of design or process problem that could be improved as well as the upper and lower operational and destruct limits for use in screen development.

2.4 THE GENERAL APPROACH TO HALT

The usual approach in running HALT is to use the step stress approach. In this approach, progressively higher stress levels of one stimulus are applied until an operational or destruct limit is noted. Once one is noted, some means of improvement is required before higher levels can be attained. Note that both design and process problems will be found, and so engineering evaluation and decisions are required as the tests progress. This fact makes it difficult, if not impossible, to lay out a detailed plan ahead of time. In their minds, this inability to define a detailed plan ahead of time is a major stumbling block to novices since the concept of a detailed test plan is synonymous with an organized approach. There is a clearly defined goal and that is to find out how to obtain increases in the product's operational and destruct levels. The indefinite part is "How much is enough?" Designers are sometimes confused by the stimulation levels used and think that they have to design for them at the beginning of the design phase. They do not. The correct approach is to use typical or expected stress levels for design and then use HALT to find the weak links in the product *with time compression*. Only the weak links are made more rugged and there is no point in ruggedizing a non-weak link since this would only increase cost and not capability or reliability. One must use logic and intelligence here so that an extreme over-design with the concomitant costs is not obtained. Some individuals have refused to use HALT because they think that one must design for the extreme stresses used. This is absolutely not true. The extreme stresses are used *only* to gain time compression.

The step stress process continues until levels well above those expected in normal environments are exceeded. During this process, continuous evaluation is done to determine how to make the unit able to withstand the ever-increasing stresses. Generally, temporary fixes are implemented just so the test can continue and more opportunities for improvement can be found. When a group of fixes is identified, a number of permanent fixes may be implemented and then new hardware built. How far to go in any one stress level is a function of too many variables to discuss in detail. The usual rule to follow is to exceed the expected environments by a comfortable margin so that *all* members of the population can be expected to survive either the field environments or the screen environments assuming that they are defect-free. If HASS is to be used, the HASS environments will be the most intense, but of limited duration, and, therefore, of limited useful life removal as will be proved in Safety of HASS.

One obvious fundamental limit usually accepted without question on most commercial products is the softening of plastics near 100°C. Softening much below 100°C is to be viewed with some alarm as this will limit temperature stressing effectiveness quite severely if some margin below the softening temperature is used. In this example, the softening plastic part would be removed from the chamber or protected from the environment by a cover through which cool air could be forced. This would allow higher temperatures on other parts of the unit under test so that additional opportunities for improvement could be found. Note that all failure modes will have a distribution, and one could be unfortunate enough to have a very robust specimen which will not readily fail under the stress conditions, but others in the population may. If the stresses are not elevated to a high enough level, some failure modes that would exist in a large sample and that would cause field failures could remain undiscovered. What is being done is to substitute stress levels for sample size. This is one of the major benefits of HALT and is a very cost-effective trade-off.

After one stimulus has been elevated and then depressed to levels felt to be sufficient, another stress is selected for step stress testing. This rotation continues until all stimuli have been applied. At some point in the process, combined environments such as temperature combined with vibration should be utilized, since there could be some interaction between stresses; that is, the combined effect may generate larger stresses than either stress alone would create. This phenomenon must be taken into account when selecting screen levels since a slight increase in stress will greatly reduce the number of cycles to failure. See Chapter 7 for further information on the physics of failure.

After all stresses are used in HALT and a number of design changes are determined to be beneficial, permanent changes are introduced, a new sample built up and then the cycle repeated until the fundamental limit of the technology is reached for all stresses separately and then in combination.

As a final step in HALT of any type, modulated excitation should be performed to allow the detection of patent defects which may remain undetectable otherwise.

2.4.1 The Protection of Fragile Elements

Sometimes an element is identified which cannot be ruggedized. There is good reason to increase the stress levels so as to find other failure modes (opportunities for improvement) which may not be apparent unless we increase the stress levels further. In order to test at higher stress levels, it is necessary to inhibit the failure mode in some manner. One way to do so is to protect the weak link by not letting the stress get to the site of the weak link. Suppose that one were doing step stressing in temperature and some component stopped working above some moderately high temperature. The way to get the product to perform above that temperature could be to isolate the critical component from the environment by placing a small perforated container over it and blowing conditioned air into the container, thereby keeping the component near the conditioning air temperature. The product could then be taken farther in temperature to discover perhaps some other failure mode which could occur in the field on a less robust member of the population. Another situation could exist and that is that a weak link is found that will cause some failures in the field, but we cannot make any changes for whatever reason and will have to live with the failure mode. It still makes sense to find other weak links and improve them.

If one were to experience some product feature that is sensitive to vibration, then the sensitive unit could be isolated for purposes of continued HALT and maybe also for field use.

It also must be borne in mind that the stresses used in HALT do not replicate those in the field and there is frequently overlap of the Venn diagrams of the various stresses. Therefore, the failure modes may not show up in the same order that they would in the field environments. So a mode that is not too likely to show up in the field may have been found, and one that is much more likely to show up in the field may not have been found. It is important not to focus on the stress which makes the weakness show up, but to focus on the failure mode instead. Much more is written on this in later chapters.

A few examples of non-relevant failures are in order:

1. A mechanical assembly showed degradation of a grease at 100°C. This is the example mentioned in Chapter 1. The manufacturer of the grease stated that the oil would slowly evaporate above 75°C leaving eventually only the soap, which would be so thick as to jam the device. The upper temperature limit on the thermal screen was limited to 80°C in consideration of this feature of the design. The design would otherwise operate properly without degradation of any kind at above 100° even

though it is intended for an office environment. This is not unusual for devices run through a proper HALT program. The temperature of 80°C was allowable due to the short duration of the screens. Proof of HASS, covered in Chapter 6, demonstrated the safety of such an approach.

2. An electromechanical device, a proximity fuse for the Sidewinder Missile, was microphonic above 20 GRMS. This is not relevant as the in-use vibration is only 1 GRMS all axes, broad band and so the microphonism would never be observed in the field. However, it was determined by tests that the microphonic signature could be used as a discriminator for finding a particular latent defect. The latent defect changed the signature in such a way that the defect could always be spotted even though the microphonism was not relevant.

2.5 IS ONE HALT ENOUGH?

In the author's experience, so much is usually found in a first HALT for any company new to the methods and unfamiliar with the types of design defects that can be designed out before any hardware is built that redesign and retest are almost always in order. The second test may (and usually does) uncover more problems and so a second redesign and/or process change iteration may be required. The retest on redesigned hardware should continue until the test results are satisfactory and the largest reasonably possible margins are attained on all relevant stresses or input variables. Again, only consider margins at the end and not while trying to determine whether any given opportunity for improvement should be exercised during HALT.

2.5.1 Step Stress Intervals Suggested

The question of reasonable step size usually comes up and is a good question. Some suggested step sizes are listed in Table 2.1, but there are many reasons for deviating from these step sizes

Table 2.1 Suggested step sizes per stress

Stress	Step size
All-axis impact vibration	25% of GRMS level
Temperature	10°C
Temperature change rate	Use max rate available.
Voltage	20% of expected margin
Power line cycling	Cycle at each voltage 20 times
Power line frequency	20% of expected margin
Humidity	10%
Other stresses as appropriate	20% of expected margin

and every case must be evaluated separately. There is no given "recipe" for doing HALT. Good engineering and financial judgment must be used throughout the tests.

These stresses should also be combined when it is appropriate to do so. Many cases have shown that thermal cycling and vibration together precipitate and detect many times more defects when combined than when applied separately. See [3] wherein it is shown that 45% of the defects were exposed *only* when an all-axis impact shaker and an ultra-high-rate thermal cycling system were used simultaneously. Many other references have shown that combined excitation is much more effective than sequential excitation by various stresses. The optimum, both in terms of effectiveness and cost, is to combine all possible stresses. However, humidity and high-rate thermal cycling are basically not compatible and other combinations also may not be in particular cases.

Note that, in HALT, a very accurate definition of the operational and destruct limits is not needed since the product will be changed to obtain the largest possible margins and the locations of the opportunities for improvements are what is needed.

2.5.2 Where to Stop Step Stress and Ruggedization

The question as to where to stop the step stress, failure mode discovery and improvement process always comes up. The novice at HALT almost always wants to stop too soon. The correct place to stop is when the cost of ruggedizing exceeds the cost of not ruggedizing; that is, marginal cost equals marginal return. Alternative uses of money must, of course, also be considered. The limit where marginal cost equals marginal return is seldom, if ever, known, so good engineering judgment is used. The fundamental limit of the technology is the goal, but sometimes the project runs out of money or runs out of time. No guidelines are given here since the breadth of such guidelines and the variation from product to product would be too broad to be of help. In addition, the author has learned that newcomers will frequently just take any number given as a guideline and use it as the absolutely best number that exists, even if it does not make sense in a given application. There is no set way to give guidelines ahead of time as one needs to consider

1. the application
2. the margins of the product
3. the correlation of design weak links found to previous products of similar construction and
4. many other factors.

These could include safety aspects, the cost of failures, the impact of a law suit due to product failure, quantity to be built, competition, or other market factors.

A suggestion, especially for the beginner, is to keep track of weaknesses found in HALT and not treated as an opportunity for improvement; that is, any that are *not* improved. If these same weaknesses cause field failures, then valuable experience will have been gained and the same mistakes will not be so likely to be repeated again. It cannot be overemphasized that beginners usually stop improvements before sufficient margins have been accomplished for a variety of reasons. It usually costs very little to obtain very large margins on most products and it may cost millions of dollars if an opportunity is missed and field failures occur. A frequently occurring cost [2] of a miss in Hewlett-Packard is US$10 000 000 per miss! That is, if a given weakness were identified in HALT and some gain in margin not obtained as a result of this discovery, then the field failures and fix implementation

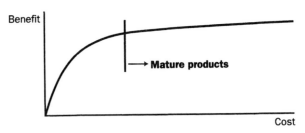

Figure 2.7 Cost–benefit relationship in HALT

usually cost a total of US$10 million dollars. Even the largest of corporations will notice losses of this magnitude.

Perhaps a graphical discussion will help determine where to stop. Figure 2.7 shows the relationship of the cost of HALT to the benefits derived. The curve drawn is for illustrative purposes only, but is seen to be of the correct shape by a cogitation on the subject. The point to be made is that continued rounds of HALT on one product will provide diminishing ROI. If other new products are ready for HALT, then it would be more beneficial to run a first HALT on them that to run an nth HALT on the product under consideration. Since companies beginning HALT are usually short of manpower and equipment for the task and cannot, therefore, simultaneously perform HALT on all projects, it makes sense to work on the project that has the largest payback on investment. That will usually be the newer product if all other considerations such as market demand and competition are equal.

2.6 AT WHAT LEVEL SHOULD WE DO HALT?

HALT can and should be done at each assembly level from the lowest assembly level to the highest. For any system, the basic philosophy is to do HALT at the lowest assembly level where it makes sense and then progress to higher levels as hardware becomes available. Using the lowest assembly levels makes sense from several points of view:

1. the lower assembly levels are usually available sooner, allowing earlier testing and more time to fix problems

2. it is frequently easier to apply very high stresses to small assemblies, and

3. the smaller subassemblies are usually much easier to monitor and to trouble shoot than is a larger system.

Sometimes, however, the lower level assemblies do not function such that meaningful tests can be performed and monitored. There is no general rule here. One just has to figure out what makes sense and then do it.

Note that subassemblies can be reused at higher assembly levels if the assembly still works normally after the HALTs done on it so far. Using all new hardware at each assembly level is not only unnecessary, it is totally cost prohibitive in most cases.

2.7 WHO SHOULD DO IT?

HALT is a complex task requiring many talents and skills for complete execution of the plan. Usually included are the following disciplines: research and development or design according to the organization in the company; advanced manufacturing (if the company has one), manufacturing, materials engineering, quality and reliability, the company chemist, the failure analysis laboratory, program management and perhaps a few more depending on the product.

The leaders in the effort should be the design teams who can most easily change the design in the best manner to increase margins when opportunities for improvement are found to exist. The other members are then in a support role. If the design team will not accept the lead role, or at least a *supportive* assistance role, then progress will be painfully slow if any progress is made at all. Assuming that the designers are at least supportive, the actual tests can be run by anyone trained in the techniques. Any engineers with experience in the field failures of the product type under consideration can add invaluable knowledge to the activity. The real talent required is that of figuring out how to exercise an opportunity for improvement; that is, change the product so that it is more robust.

2.8 WHERE SHOULD HALT BE DONE?

Considering the staff required to be on call during HALT, it makes sense to do the tests in-house or very close by. The performance of HALT at a remote facility is somewhat expensive when travel is considered. The tests become very expensive if not properly supported because of the costs of travel and the many repeats necessary due to the poor support. Rental or leasing of the proper equipment is generally much cheaper and much more effective than traveling to some remote site with all of the staff and test gear required in order to perform the tests properly.

Still, many companies choose to verify that the methods work on their products before any such expensive step. At first, they use one of the labs staffed with knowledgeable people and having the correct equipment for HALT. The author calls this stage the proof of concept stage. It seems that management is usually reluctant to buy, or rent (with a 40% of purchase price commitment plus installation charges), equipment of the types required for proper HALT and HASS unless it is proven that HALT will work on the company's products. Of course it does, but management does not know this for sure, partially due to the lack of state-of-the-art published papers. The proof of concept stage is a waste of time and money in retrospect, but managers are just not willing to spend the money unless definitive tests have been done on their own equipment lines. This is one of the many reasons why most of the leaders do not publish complete success stories in this field. Lack of successful implementation results from other sources slows the implementation by a few years in most cases. After struggling through the acceptance of the idea of time compression, the proof of concept, the justification for purchasing equipment and then performing a successful HALT and HASS program, one does not want to give away the expensive and slowly learned lessons to the competition. For this reason; that is, the lack of published data and methods, many current HALT and HASS papers are using the technology as it existed in 1986 and as published in [5] and [6] in 1991. These references were intentionally old technology and very abbreviated so that the author could maintain his own lead in the technology.

Two suggestions are made:

1. For the proof of concept tests, use a product which is known to have problems, therefore assuring that it is possible to find something. This approach proves, without a doubt, that HALT picks out relevant defects very quickly, but, unfortunately, the damage is already done in releasing an immature product. The test results are difficult to ignore if, as is normally the case, the same defects show up in HALT as have been showing up in the field.

2. Use a brand new product that is in the development stage so that the maximum benefit is gained from the tests. A drawback in this approach is that the designers may refuse to fix anything because the product was stressed beyond the field levels. Choosing to ignore or explain away such failures, as explained several times before, is one of the most frequent errors made by a novice. This is also typical of designers when first exposed to HALT.

Running both types of product seems to be a good compromise situation, convinces the skeptics that HALT really does work and also gains the maximum benefit of HALT on the new product.

It is suggested that the newcomer avoids the pitfall of taking a mature product as the sample in a proof of concept test. If this is done, then probably nothing will be discovered. The conclusion will be drawn that the techniques do not work. In reality, it is the fact that the product was mature that led to the result that nothing was found as an opportunity for improvement. The author has seen many companies make this mistake.

A third suggestion is to have an in-house seminar on the subject in order to gain support for the effort. The methods that are obviously correct once one has been educated in them also seem obviously incorrect to those who have not been educated in them. The seminar can educate the uninitiated so that the methods are accepted, at least for a proof of concept trial. Since the trial always demonstrates that HALT does indeed work as claimed if done correctly, then it represents a major milestone in progress toward initiation of the techniques in the plant.

2.9 HOW MANY UNITS SHOULD WE HALT AND FOR WHAT ELSE CAN WE USE THEM?

The question of how many units to HALT is one that has an obvious answer, "As many as possible." The next relevant question is – "How many are possible?"

During early prototype fabrication, few units will be available but the search for problems will be fruitful and many problems will be found. These problems will soon swamp the design people, at which time there is little point in identifying more problems at this stage. When the identified problems have been fixed, it is then time to fabricate improved hardware and to do some more HALT. How many to test is to a large degree affected by what we do with the units afterward; i.e. are they throw-aways? It seems that there are several uses for units used in HALT. Some of these are:

1. Use them as Safety of HASS units. The author has always succeeded in having HALT units pass Safety of HASS. This is not going to always be true, but it is particularly worth trying if the units are very expensive.

2. Use them for engineering change modeling and then HALT again to ascertain whether progress is being made to attain more robust hardware.

3. Use them for internal tasks such as test equipment development. It seems that at least a few test units are necessary on each program and refurbished HALT units will fill this need very well.

4. Supply them to Original Equipment Manufacturers (OEMs) as engineering evaluation units. OEMs usually just want to verify form, fit and function, so HALT units will suffice for this need. It is wise to tell them that the units have been subjected to HALT so that they understand just what it is that they are getting and to what environments the units have been exposed and that the units may reach end of life at any time. Generally, HALT units become "Golden Units"; that is, units which can be counted on to perform flawlessly.

5. Give them to marketing for shows, demos, etc. One could even briefly note the history of each one to demonstrate the company's commitment to quality and reliability. Trade show attendees are usually very impressed at the abuse that the product can take and still come up working, which leads to sales.

6. Sell (or give) them as HALT units at a reduced price to beta sites. A functioning unit being used by a customer for an extended duration gives very valuable data on many aspects of the unit. The author has been involved in selling HALT units on two occasions. In both cases, no failures occurred during the (extended) warranty period. Both parties were delighted with the transaction. The buyer got a well-screened unit quickly and at a very reduced price and also got an extended warranty. The seller got some of his development money back after getting all of the information that he sought. This can be a win–win situation if everyone involved understands what HALT is all about. The author regularly looks for HALT units for his own use when purchasing something that may have been through HALT. This does not apply for mission critical items, e.g. an artificial horizon instrument which is critical for instrument flight in an airplane.

With all the uses above, and perhaps more, it seems that one can afford to do plenty of HALTs if everything made in prototype production is run through HALT. One should HALT every serial number at each assembly level until production starts, and production should only start when we are satisfied with the HALT results. Note that the demand for working units for various puposes such as software development, sales, etc. can usually be satisfied with HALT units. This approach requires a large break with the old and obsolete tradition for most companies, but the results are:

1. much faster design and process maturation
2. earlier (mature) product introduction
3. reduced total engineering time

4. reduced total costs
5. fewer field failures
6. higher profits
7. increased market share.

2.10 REPEATED HALT (RE-HALT)

2.10.1 Verifying Design Capability

Some means of verifying that the design remains as capable as when the original HALTs were run should be repeated on a regular basis during production. It is surprising how many design and process changes can creep into a production product in just a brief period of time. Re-HALT is usually run only at the corners of the stimulus envelope; that is, only at the operational and destruct limits of each stimulus separately or at other critical sets of parameters as determined in the last series of tests. For example, if low voltage and high frequency power have been found to be a critical set of parameters, then this combination should be explored on every retest. This approach ensures that the product is as capable as before and saves test time. If capability has declined, then some form of corrective action is usually required. The first impulse of the novice when finding a change in capability is to say that the tests are destructive because they exceed the design qualification limits (the same argument is given by the novice when seemingly good hardware fails in a screen). *There is always a reason for a decline in capability, and that reason should be sought out right down to the root cause.* Correction is usually called for, but is sometimes not possible. In some cases, the decline in capability is completely beyond control (such as in purchased parts or subassemblies where there is only one vendor) and the new operational and destruct limits are noted and the screens are accordingly adjusted until corrective action can be implemented or some means of protecting the new weak link is determined. The impulse to reduce screen intensity should be resisted since continued application of this tendency would reduce the screens to mere "show screens" which appear to be screens, but in reality will find only very gross defects. These "show screens" have most of the costs but few of the benefits of proper screens. If a product has passed Safety of HASS in the past, then if failures are occurring in the HASS, something has changed for the worse. The capability should be recovered if possible.

2.10.2 Verifying Margins

The large margins obtained in the design ruggedization process will not automatically remain throughout time, but will decay as design and process changes take place, sometimes without any knowledge of that fact. The HASS profiles proven safe (Chapter 4) will only remain safe if the large margins remain in place. For that reason, one must re-verify that the margins remain at least as large as they were when Safety of HASS was performed. This is accomplished by again performing HALT and determining that all margins are as large as they were when Safety of HASS was performed. Chapter 4 will show that the margins must remain at least as large as when Safety of HASS was performed in order for the proof to remain valid. If it is not valid, then many field failures can be expected due to life removal during HASS.

There are several occurrences that should trigger a re-HALT including: starting to fabricate on an additional shift; opening a new manufacturing plant; going off-shore for manufacturing; or obtaining subassemblies from new vendors. Any of the above occurrences can affect margins to a large degree. Particularly when manufacturing off-shore, one should perform a HALT on the product built off-shore. Note that many changes may occur when going off-shore including: a new dimensional system; new tooling; new language; new vendors; a different set of atmospheric and pollution conditions; and a new culture. Sometimes it makes a difference if the assembler is right- or left-handed!

The existence of the required margins is easily evaluated using a simple re-HALT. The step stressing need not be begun at the lowest level in this case as only the margins need to be determined. Therefore, the step stress testing can begin at levels near those attained in the original HALT, the one used to establish the HASS levels. It may be sufficient to just repeat the combined stress finale to a usual HALT sequence in order to demonstrate that the margins are as they were before.

If the margins have decayed, then either the margins must be recovered or the HASS profiles must be adjusted. A HASS profile that has been proven safe for a product is valid only if the margins do not decrease. If the margins have decreased substantially, then the HASS can become degrading or even destructive to the hardware in extreme cases. The author has seen this occur several times over the years and the results can be disastrous to the finances of the producing company.

2.11 WHY A PARADIGM CHANGE IS REQUIRED TO DO SUCCESSFUL HALT

An attitude of wanting to find problems is required in the program outlined, whereas many engineers have been trained, or are motivated, to "pass the test". The "pass the test" philosophy, so prevalent on military contracts in years past, usually means that one carefully hand-built unit is run through the Qual tests with the *goal of passing*. Once one unit has passed, the design is frozen and production is begun. This approach is certainly different from HALT where everything imaginable is done in *order to force* failures and then to fix the source of the failures. A few quotes (all misguided) from the author's experiences may illuminate the points to be made:

"There can't be a problem, we've always done it this way!"

"But it only happens below −20°C and spec is 0°C!"

"No problem, it only happened in one out of ten!"

"It's only a random failure!"

"It's just an infant mortality!"

"Don't worry, it's only a process problem!"

"No problem, we'll fix that after manufacturing release!"

What the designers frequently say is:

"Well, it wasn't a hard failure!"

"Of course it broke, it wasn't designed for that!"

"Where will this product ever see that kind of stress?"

"Of course it failed, you took it over spec!"

The last three are some of the most difficult misconceptions to overcome since many have it in their heads that anything "over spec" is unreal, unwarranted and just plain stupid! This attitude displays ignorance of time compression. Experience with proper HALT and HASS techniques since 1969 has shown that stresses up to the fundamental limit of the design produce *relevant* failures, almost without exception. As stated before, any design or process weakness not improved will likely show up in the field. A guest speaker at a recent seminar said that a particular failure mode was found at 80°C in one of his products. The design team refused to improve the design because the failure was found 20°C out of spec. The speaker went on to relate that 85% of the field failures were that exact mode! If you find it in HALT, it is most likely relevant!

Proper application of the techniques requires an attitude of wanting to improve the product to the fundamental limit of the technology any way that it can be accomplished. We must discard the philosophy of "If it is not broken, then don't fix it!" and replace it with "If it is not perfect, then we'll make it better until it is as close to perfect as we can make it!" Properly designed and built equipment today can result in products that virtually never fail during a normal lifetime of field use.

2.11.1 Failure Modes and Their Relation to Stress

It is extremely important to focus on the failure mode and mechanism and not on the stimulus used to find the weak link. Often, one finds a weak link using some stimulus that is different from the one that would cause the failure to occur in the field environment. This is because various modes of failure are accelerated by different amounts when step stress testing is done. For example, if a certain mode is not accelerated much by thermal cycling at broader ranges or at higher rates, it may be accelerated by a very large amount in vibration. Even though vibration may not cause the field failure, it may uncover the weak link in accelerated testing. To ignore the weak link because it was found "out of spec" is a major error.

2.12 TIME AND COST SAVINGS FROM HALT

The HALT process takes place during the earliest possible design phases so that prevention costs can accrue. Let us compare a typical, but now obsolete, approach to product design, and the HALT approach. First, let us consider the old way. The design is begun after staffing up for the effort. The costs run as shown in Figure 2.8 which illustrates the spending rate as a function of time. Multiple tries at a Design Verification Test (DVT) are made and only a few problems are found during each test since only field level stresses are applied, resulting in only very gross defects being found. Finally, after several tries at a DVT, a successful one is performed, the design is frozen and production is begun (Manufacturing Release, MR, on Figure 2.8). At this point, the fate of the product's cost and reliability are nearly determined with 85% of total costs now essentially fixed by the design. Note that, in this

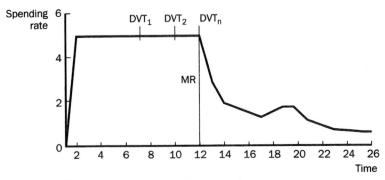

Figure 2.8 "Classical" approach spending rate

case, an immature design has been released to production. As a result, there will be many field failures since all design robustness limiters have not been found and a significant cost of sustaining engineering will be required to nurture the low-margin product thorough its production life cycle. Design problems or robustness limiters are frequently not fixed after product release to manufacturing for many reasons. Among them are: compatibility of parts and subassemblies; the inability or reluctance to recognize a design problem in something that passed "Qual"; sunk costs in tooling (one should never base decisions on sunk costs, however); the inability to accept that the design is less than ideal; the need for blame-placing if a design fault is found or denial. Note that sustaining engineering and field repairs go on and on at a high activity level after product release to production. There may even be a factory recall or two due to the release of the immature product.

With the HALT approach, illustrated in Figure 2.9, overstress testing begins as soon as there are prototype units available for testing and so testing begins much sooner than in the classical approach. Problems (robustness limiters or opportunities for improvement) are found and fixed very rapidly by the approach since severe overstress resulting in time compression is used to force the weak links in the design and processes to show up quickly. Since the design is still evolving, changes are easily implemented in the design. During the HALT phase of product design, there is very intense activity on the part of the design engineers, the test lab, the technicians who troubleshoot and repair

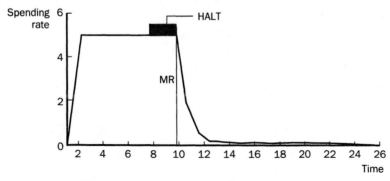

Figure 2.9 HALT approach spending rate

the product, the failure analysis group, project management, purchasing and other groups or individuals. The spending rate is somewhat higher at this point in the design phase than if no HALT were to be done. After the fundamental limits of the technology have been attained, the product may be run through a DVT if management insists. Most experienced companies delete this time-consuming and expensive test since all stresses have already been applied far in excess of the "spec" range and it is almost certain that nothing will be found at the lower stress levels if appropriate detection screens have been run during HALT. Some defects can only be detected with modulated excitation (see Chapter 4). Newcomers to the techniques will probably insist on doing the Qual tests or verification tests the first few times that HALT is utilized. The current thinking may be "We have always done Qual tests!" and somehow they think that a design is not completed until the Qual test is passed. A Qual test is usually a non-event after a HALT! This is another benefit of a properly done HALT; namely, deletion of major portions of the Qual tests.

Once the HALT is completed and the design and processes are brought to maturity, production of a mature product can begin. Since the product is fully matured during the HALT process, sustaining engineering and warranty work will be minimal. The costs will therefore drop off as shown in Figure 2.9. Figure 2.10 shows a comparison of the two approaches. The saving is the crosshatched area for sustaining engineering less the shaded area during HALT. These areas representing costs are usually very substantial.

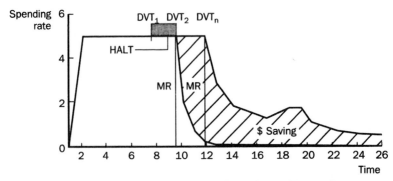

Figure 2.10 Comparison of "Classical" and HALT spending rates

On some occasions, the time to release with HALT has exceeded that to release without HALT, but, in these cases, the product had numerous design weak links and release of such a product would have been sure to lead to many expensive field failures.
Using the HALT approach obtains the following:

1. a mature product will be attained much sooner
2. production release will usually be sooner
3. sustaining engineering will usually be much less
4. warranty costs will usually be substantially less
5. customer satisfaction will definitely be much greater
6. greater market share
7. economies of scale
8. higher margins, etc.

The author has heard some reports of products missing their market window because HALT exposed so many problems that they could not be eliminated before the window was missed. This is really a success for HALT since, in all cases reported under non-disclosure agreements, a sure financial disaster would have occurred had such a poor product been released to mass production. Two companies reported that bankruptcy would have occurred if the products had been released to production before the conclusion of HALT.

2.13 BECOMING "PRODUCT SMART"

It is very normal to find many problems in a new design using these methods. More often than not, however, it is found that just a few weak links exist in the product, and, when they are removed, a very robust product is at hand. Gaining the experience of eliminating defects leads to knowledge as to how to design them out in the first place. This is a major long-term benefit of HALT. After individuals, and then collectively their companies, have experienced HALT a few times, design rules for the product line will be established and then HALT will proceed very quickly with very few discovered opportunities for improvement. At this point, the individuals and company have become what the author calls "design smart". This status usually takes a few product cycles in very progressive companies.

In one company, which shall remain unnamed for obvious reasons, it took about seven years for the design engineers to completely accept HALT, to also accept the lessons learned from previous HALTs and then to design from an enlightened point of view. After the learning period had passed, HALT generally exposed nothing that represented an opportunity for improvement, HASS exposed virtually nothing and so progression to HASA was made. In spite of the rather slow progress, the company reported informally that "HALT and HASS saved the company hundreds of millions of (US) dollars (in the first two years)!" Imagine what would have happened if the company had enthusiastically embraced the methods immediately. The company moved in the first two years of HALT and HASS from fourth place in their industry in terms of field reliability to first place in spite of the slow acceptance.

In companies where HALT was accepted readily, the progression to "design smart" took only a few product cycles, which represented only six months or so. In these cases, the cost savings accrued much more rapidly, as did the reliability.

2.14 THE BIGGEST RESISTANCES TO ACCEPTING HALT

Many people exposed to the HALT philosophy for the first time resist the concept of time compression in spite of the fact that there is solid theoretical support and engineering evidence of the effectiveness of the concept. The author has found over many years of performing seminars on the subjects of HALT and HASS that proper training on the methodology alleviates the resistance to the new philosophy and speeds acceptance and implementation substantially. Unfortunately, many published papers on the subject reveal less than complete knowledge of the methods, resulting in less than optimum results in many cases and total disasters in others.

A second misconception is often espoused by the same people who have the "over spec" syndrome. They often claim that more money is spent building a rugged product than a non-rugged product, one that just meets "spec". This simplistic view may be true only if a very narrow definition of cost is taken. If one were to consider only the cost of design and production (not counting rework), then it may be cheaper not to perform HALT and HASS. More accurately, the total costs over the product's lifetime should be examined. These include: screening, rework, inspection, tests, warranty, the impact of lack of customer satisfaction and market share loss. All of these costs, and even more not mentioned, should be accounted for when making decisions on whether or not to perform HALT and/or HASS.

A major failing of many in management situations is the "short-term mentality" syndrome. Short-term mentality concentrates on this week's production, this quarter's profit, expenses today, and so on. The author has even observed companies that give production managers bonuses based only on the total number of products shipped without any consideration of their quality or reliability. In this case, the manager will be encouraged to ship anything that can pass the required tests (if there are any). Predictably, the manager would not accept HASS as it would slow down production by exposing flaws which would have to be fixed and this would decrease his quarterly bonus. The design manager also had a similar bonus plan based on how fast he could turn out a design that would pass "spec". He would not accept HALT as he viewed it as one more task to slow down his progress and reduce his bonus. Effective HALT and HASS requires long-term, life cycle management thinking. Those companies that have such management have done very well indeed with their HALT and HASS programs.

A third misconception is present in the following attempted rebuttal made by skeptics: "If HALT is so great, then why is not everybody doing it?" Many are doing it, they are just not broadcasting their results. ROIs on HALT activities range from 10:1 up to 1000:1 or even more occasionally. It is no wonder that the leaders will not give away their technical lead since it is extremely difficult to get ROIs of this magnitude any other way! As a point in fact, the author has signed more than 200 confidentiality agreements regarding HALT activities and most will not allow the disclosure of even the client's name, the product, the fact that the client had an in-house seminar or that the consultant was even there for any purpose at all.

2.15 MTBF

In the process of finding the upper and lower operational and destruct limits and pushing them to the fundamental limit of the technology or within the bounds of time and/or money, a *very* robust product of far superior, that is, low, field failure rates will be generated. This aspect of HALT is the major benefit of the method. The product is made better, but how much better is not known, at least not by the HALT methods. However, when results of the MTBF based on field failure rates (the only meaningful MTBF) become known, it will probably be far higher than ever before on this type of product when HALT is done for the first time. One must use the usual expensive and time-consuming techniques in order to determine the MTBF if an estimate of that parameter is required. The author considers the MTBF to be an inappropriate parameter to consider in the first place because, if the type of failures which are going to occur are known, then one should use the available funds to improve the weak links instead of trying to figure out how often the product will fail and spending a lot of money to do so. Note that HALT pushes the bottom of the bathtub curve down and the wearout portion far to the right, but no information as to the height of the bottom of the curve is obtained. If the bottom is not at essentially zero, one should be working on getting it there instead of finding out exactly where it is.

To quote Larry Edson of the Cadillac Luxury Car Division of GM in 1998, "Legendary reliability is possible after a few short HALTs. The problem is that the measurement of this reliability is darn near impossible." In addition, Mr Edson related that the HALTs cost only 10% of the cost of running the long, expensive MTBF test. In

MTBF   73

this case of a tail light assembly, it had been determined that the MTBF was 55 car lifetimes long. Getting excellent reliability was cheap, measuring the MTBF was much more expensive! This situation is characteristic of the results of a proper HALT program. It seems that obtaining very large margins on ten products is much more valuable than obtaining the numerical value of the very large MTBF on only one product.

2.16 SUMMARY

The methods of performing HALT have been discussed. Any stress or other parameter that exposes opportunities for improvement is suitable in HALT. It is always prudent to perform modulated excitation to put a flaw into a detectable state after it is patent. A block diagram of the HALT process is shown overleaf in Figure 2.11.

Some features of HALT are:

1. HALT is an information gathering tool that allows the design and fabrication of products with very large margins between the expected field environments and the capability of the product with only minor cost increases during the early design stage.
2. HALT is a modern design approach and results in savings in design, fabrication and warranty.
3. HALT does not lead to the prediction of field failure rates, but it does significantly reduce them.
4. HALT forces product maturity and speeds product introduction as well as lowering overall development costs.
5. HALT will rapidly find almost all of the same failure mechanisms as the field environments would if HALT were not done.
6. HALT helps improve the product to the fundamental limit of the technology during the design phase.
7. HALT will help a company and its employees to become

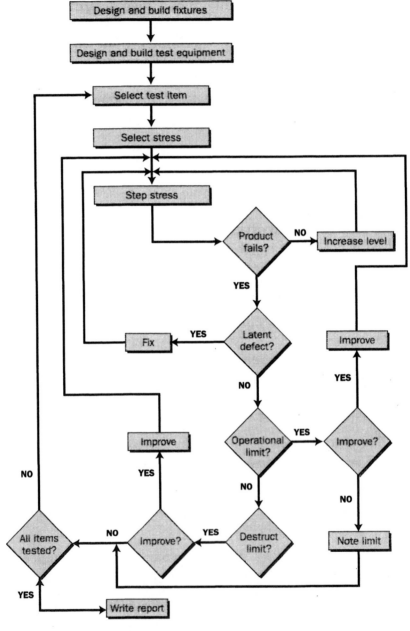

Figure 2.11 The HALT process

"design smart" and, therefore, to design products which do not have weak links.

8. HALT requires a cultural change for most companies not familiar with the technique.

REFERENCES

1. Edson, L., Combining team spirit and statistical tools with the HALT process, *Proceedings of the 1996 Accelerated Reliability Technology Symposium*, Hobbs Engineering Corporation, Denver, CO, 16–20 September 1996.
2. Seusey, C., Achieving phenomenal reliability growth, *ASM Conference on Reliability – Key to Industrial Success*, Los Angeles, CA, 24–26 March 1987.
3. Silverman, M., Summary of HALT and HASS results at an accelerated reliability test center, contained in the HALT and HASS seminar notes of Hobbs Engineering Corporation. This paper is available upon request without charge.
4. Hakim, E. B., Microelectronic reliability/temperature independence, US Army LABCOM, *Quality and Reliability Engineering*, 7, pp. 215–220.
5. Hobbs, G. K., Highly accelerated life tests – HALT, *Proceedings of the 1992 IES Annual Technical Meeting*, Nashville, TN, 3–8 May. Although published in 1992, this paper was based on 1986 technology and was very abbreviated.
6. Hobbs, G. K., Highly accelerated stress screening – HASS, *Proceedings of the 1992 IES Annual Technical Meeting*, Nashville, TN, 3–8 May. Although published in 1992, this paper was based on 1986 technology and was very abbreviated.

CHAPTER 3
Hass: Highly Accelerated Stress Screens

A new scientific truth does not triumph by convincing its opponents and making them see the light, but rather because its opponents eventually die and a new generation grows up that is familiar with it.

Max Planck, Scientific Autobiography

3.1 INTRODUCTION

Highly Accelerated Stress Screens (HASS) are used in the screening of production items on a 100% basis using stresses which are substantially higher than those experienced in normal use including shipping and storage. The technique may also include stresses which do not occur in expected use if these stresses help locate flaws which would occur in an expected field environment. The stresses not occurring in the normal environment are used due to the overlap of the Venn diagrams of flaw/precipitation relationships and due to the fact that some precipitated defects are only detectable when exposed to certain environments or combination of environments. Chapter 7 describes the relationships between the stresses used and the flaws found. In HALT and HASS, the only two requirements placed on the environments (or other stimuli or methods) used are that they

1. precipitate and/or support the detection of relevant defects, i.e. the family of defects which would show up in normal use including shipping and storage, and

2. also leave the products with substantially more fatigue life left after screening than is required for survival in the expected field environments. (Note that this definition of screening is quite different from that used for classical environmental stress screening.)

This chapter discusses some of the underlying concepts of HASS and goes into how to actually select the regimens based on the operational and destruct limits determined in HALT. Chapter 4 describes in more detail how to show that the selected HASS profiles do not remove too much life from the product while actually precipitating the latent defect to patent and then detecting the patent defects. Chapter 5 discusses screen optimization for minimum duration and cost. It is noted that some of the techniques discussed in the present chapter are used during HALT, so this chapter should be read and understood before attempting HALT.

3.1.1 Precipitation and Detection Screens

A screen is intended to detect latent defects that would cause a failure during the manufacturing processes or in the field environments, including transportation, storage and use. Correctly done, screening is a closed loop six-step process consisting of:

1. *Precipitation* This means changing some flaw in the product from latent (undeveloped or dormant) to patent (evident or detectable). An example would be to break a nicked lead on a component or to fracture a defective bond or solder joint.

2. *Detection* This means observing in some manner that an abnormality exists, either electrically, visually or by any other means. In the cases illustrated above, one could visually or electrically detect that a lead, a bond or joint had broken.

3. *Failure analysis* This means determining the origin or cause of the flaw. In the illustrations above, we would determine where in the production process and why the lead had been nicked, why the bond had not been properly made or why the solder joint had not been properly made.

4. *Corrective Action* This means implementing a change intended to eliminate the source of the flaw in future production. The nicked lead might be prevented by using a correct forming die, the bond might be corrected by using a different pressure or temperature or perhaps better cleaning and the solder joint might be corrected by using a different solder or a different temperature.

5. *Corrective Action Verification* This means demonstrating that the corrective action taken actually has removed generation of the flaw type which is under investigation. Verification generally requires a repeat of the conditions under which the flaw was detected to ascertain that the anomalous behavior is no longer present.

6. *Documentation of knowledge gained* This is information gathered in a database to prevent making the same mistake again. This last step seems not to work very well due to the fact that corporate knowledge exists in many minds, people change companies, people retire, etc. Still, it is worth doing for the gains that it does make available. Through this effort, a company can become "design smart" and "process smart" meaning that the employees know how to design products and select processes based on past successes and failures. This end goal of becoming "smart" is one of the greatest benefits of any reliability program.

Every one of the five primary steps in conjunction with the others is necessary for a comprehensive screening program. Any less than the first five will not suffice to provide a truly successful screening program with all of the attendant benefits that all five rigorously done would provide.

   Precipitation and detection screens were originated and used by the author in 1979. Moderate stress, above the field environments, was used for precipitation and then a preliminary form of detection screen called "tickle vibration" combined with low temperature was used for detection. This technique was far superior to the previously used methods and so they were introduced and taught in seminars beginning in 1981. A few years later, it was found that screens could be much shorter and less expensive if even more severe stress were used for precipitation, reducing time and obtaining more cost compression. After precipitation, much lower stress levels were used for detection during combined thermal and vibration excitation. In the early 1980s, vibration was usually done only at high and low temperatures. It was subsequently found that many flaws were only detectable at intermediate temperatures and at various vibration levels. Some were detectable only during temperature change. Improvements have continued to be made in this still emerging technology and a recent development, fully modulated excitation, conceived in early 1996, has increased the effectiveness of HALT and HASS by orders of magnitude. Note that detection screens are a very important part of the discovery of defects in HALT, and a HALT done without detection screens, or, worse yet, without any monitoring at all, is very misleading. Many failures can be present and not detected if the stimuli do not put the flaw into a discoverable state and/or if the software cannot detect the flaw even if it is present and detectable. These facts are two of the reasons that users not well informed in proper HALT and HASS techniques and attempting to perform them find little to fix. These users still have high field failure rates on equipment which has supposedly "been through HALT and HASS". It is quite essential to understand the subtleties and these will be discussed in this chapter.

There is a great difference between precipitation and detection screens, yet little is found in the literature on the difference [1]. A clear understanding of the difference will allow one to develop screens that take advantage of this fact. Precipitation screens can be run beyond operational limits of the equipment (if proven safe in terms of fatigue damage) in order to accelerate the process and then detection screens can be used to detect the precipitated defects.

3.2 THE OBJECTIVES OF HASS

Some of the objectives of HASS are to:

1. precipitate relevant defects from latent to patent at minimum cost and in minimum time;

SELECTING THE STRESS MAGNITUDES 81

2. detect as many defects as possible at minimum total cost and in minimum time in order to reduce feedback delay and cost (a reduction in the feedback time will reduce the number of defective units manufactured and needing to be repaired, or maybe discarded, before corrective action is taken; a delay in corrective action will lead to the shipment of many defective units before the corrective action is accomplished;
3. provide the start of a closed loop failure analysis and corrective action program for all defects found in the screens;
4. increase field reliability by reducing the total number of faults sent to the field;
5. decrease the total cost of production, screening, maintenance and warranty;
6. increase customer satisfaction substantially;
7. gain market share through customer satisfaction;
8. improve profits via economy of scale due to higher volume.

3.3 SELECTING THE STRESS MAGNITUDES

In HALT, various operational and destruct levels were discussed. These are not just one value but are distributions over the population. The distribution is usually not known with any accuracy, or perhaps at all, and so generous margins are generated between the destruct level as measured in HALT and the HASS levels used in production screening.

Since the various levels will vary with time as the production processes and even the design change, the non-destruction part of Proof of HASS (covered in Chapter 4) should be re-verified on a regular basis. Alternatively, and much more cost effectively, HALT can be re-performed from time to time to verify that the critical operational and destruct limits of the device have not changed.

The higher the stress used, the faster the defectives will incur damage, such as fatigue, wear, electromigration, and the faster the non-defectives will accumulate damage as well. Equation 1.1 in Chapter 1 shows that the more highly stressed areas will accumulate damage much faster than the lower stressed areas. This means that the defective areas will fatigue to the point of failure while only minute amounts of damage will be done to the defect-free areas. Whether accelerated techniques are used or not, by the time the defects are precipitated, nearly the same amount of damage to the defect-free areas will be done in either case. To clarify, HASS generates nearly the same ratio of damage for any given failure mode at flawed and at flaw-free sites as would the simulation of the field environments. HASS does it with substantial time compression, which leads to cost compression.

The procedure in HASS is to use the highest possible stress in order to compress the screens as much as possible and to gain time compression. This leads to less test equipment, fewer test technicians and less power and liquid nitrogen consumed.

3.4 AN EXAMPLE OF THE RESULTS OF HASS ALONE

An electromechanical product at Storage Technology Corporation had been in production for about eight months when 100% HASS was begun without the benefit of HALT. Figure 3.1 shows the Field Replaceable Unit (FRU) life in months as a function of time in the field. Both the first and second quarter's production (unscreened

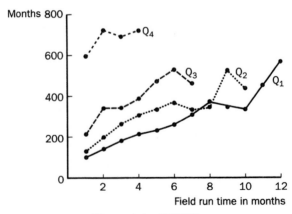

Figure 3.1 FRU life

and without the benefits of HALT) showed a growing FRU life as time in the field accrued, flaws caused failures and repairs were made leaving fewer flaws. This behavior indicates a product which could benefit from screening. Note that the second quarter's production showed reliability growth due to corrective action based on what failed in the first quarter's production. In the last few weeks of the third quarter, 100% HASS was begun without the benefits of HALT. In order to set up the HASS profiles, tests were performed to find the operational and destruct limits (this is not HALT as no improvements were made), the screens were selected as suggested in this book, Safety of HASS was performed and then HASS tuning was completed. The fourth quarter's production FRU life increased by a factor of 4.7 times compared with the last full quarter's production without HASS, which had been the second quarter.

The Pareto chart for field failures for the first two quarters of production matched the HASS failures from the fourth quarter's production exactly! This is confirmation of excellent screens. Rarely is the match exact, but it is usually very close. The failures would have been fewer if HALT had been performed first, but the author had no choice in the matter.

3.5 SCREENING EFFECT ON RELIABLILTY

When no screening is done, the failure rate in the factory and in the field may be as shown in Figure 3.2. The failure rate in the field will be too high, as shown. If screening is begun without corrective

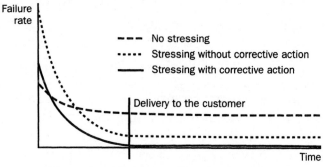

Figure 3.2 Screening effect on reliability

action, the curve may be as shown with many of the flaws caught during the production screens and with many fewer escaping to the field than before, but with still too many failures occurring in the field. If corrective action is implemented, then the number of flaws extant in the product will decrease with time and the failure curve will approach the one shown. The bottom of the bathtub curve should approach zero. If it does not, then something could be improved and corrective action should be undertaken until the curve declines to zero. Today's electronics should have an inherent failure rate of zero and some indeed do! See [2] for many papers on this subject.

3.6 PRODUCT RESPONSE TO THE STIMULI

In screening, either in the design phase looking for opportunities for improvement, HALT, or in the production phase looking for process defects, HASS, the important factor to control is product response to the stimuli, not the input. For example, in vibration screening, it is not at all important what the shaker motion is but what the product response to the stimulus is. Modal excitation to a level high enough to precipitate defects is all important in screening and so all modes must be excited. It is especially important to excite the lowest frequency ones as these usually generate the highest curvature of the circuit board, resulting in the highest solder joint stresses. Excitation of the lowest frequency resonances calls for the best six-axis shakers as described in Chapter 9. Note that the usual MIL-SPEC approach called for a very tight control of the shaker input power spectral density (PSD) and overall GRMS level and did not address the product response, completely backward from what is needed for a good screen. Another example comes from thermal cycling where the product temperatures and differences in temperature are the important variables. Close control of ramp rate, chamber temperature and uniformity of chamber temperature are just not important in this application. Again, the MIL-SPEC approaches called for very accurate control of ramp rate, temperature and the variation of temperature within the chamber. None of those variables need to be accurately controlled for effective screening. See Chapter 6 for a discussion leading to this conclusion. These examples should make it quite apparent why the MIL-SPEC approaches to screening were so costly and ineffective and why some of the most recent improvements in screening equipment are in types of products not having the ability to control accurately the input stimulus but are able to enhance the effect of the stimulus on the product. Specifically referred to are the all-axis shakers and the extremely high-rate thermal chambers now available at costs of much less than the equipment required to perform the less effective MIL-SPEC screens. A paradigm shift in equipment used for screening, as well as techniques, has taken place. The US military is now calling for the best of commercial practices including HALT and HASS. This began happening in the mid-1990s and is now well under way. Typically, the US military now requires warranties in the 10–20 year range.

3.7 SELECTION OF PRECIPITATION SCREENS

A precipitation screen is intended to convert a relevant defect from latent to patent in the least possible time and, therefore, at the least possible expense. Precipitation screens tend to be more stressful than detection screens. An example of a precipitation screen would be high-level all-axis broad-band vibration which accumulates fatigue damage rapidly, particularly in areas at, or near, a flaw, where stress concentrations usually exist. Another example of a precipitation screen would be high-rate, broad-range thermal cycling which is intended to create low cycle fatigue in the most highly stressed areas, which, fortunately, are usually found at, or near, a flaw. A third example is power on–off switching which is intended to generate electromigration at areas of high current density, usually at a flaw. A fourth is to generate rapid temperature changes which force low cycle fatigue in areas of high stress, usually near a flaw. Combined stresses are generally the most effective in precipitation since the cumulative stresses are generally higher, but not always. There are some specific flaw types located in a particular geometry that respond better to only one stress. This is not typical, however.

In using HASS, one uses the highest possible stresses which will leave non-defective hardware with a comfortable margin of fatigue life above that damage which would be done by remaining screens and the shipping and in-use environments. This approach really demands the application of HALT techniques and design ruggedization as discussed in Chapter 2 on HALT in order to be able to rapidly and cost-effectively precipitate flaws. Without using HALT techniques, the application of HASS is usually not possible due to system design fragility levels existing below the applied stress level leading to the failure of non-defective items. "Non-defective" as used here means those items which would survive a normal field environment if no stress screening were done. The precipitation level is illustrated in Figure 3.3 below. The initial value of the stress is picked using the data gained in HALT on the product and in screening similar devices. It should be recognized that the stress level picked is a first estimate of an appropriate stress level and that the final HASS profile is very insensitive to the initial estimate. This value of stress will be modified if necessary during Proof of HASS covered in Chapter 4 and/or HASS optimization covered in Chapter 5. Only the most restrictive distributions are shown in Figure 3.3 for clarity. Many exist.

Note that precipitation screens may well be run at above (or below) a design upper (or lower) operational limit where the system cannot perform and therefore cannot be tested. The use of such high stress levels is done to obtain time compression and therefore cost compression. Since the product under test is unable to perform normally during the stressing above the operational limit, one would not be able to detect completely an anomaly

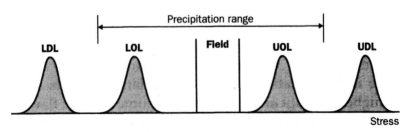

Figure 3.3 Selection of precipitation screens

during the screen. Partial functioning and detection may exist in some cases. If a precipitation screen were performed without monitoring and then the product tested on the bench under ambient conditions, many of the precipitated defects would remain undetected. Experience has shown that the percentage undetected is very high, ranging up to 100% in many cases. This is where the detection screen comes in. It is noted in passing that many current published papers are based on screens without monitoring during stressing or using only one stress at a time. It is fully expected that most of the flaws, even if precipitated, will not be detected in that case. Precipitated flaws which are not detected in the screens will almost certainly be discovered by the customer, usually in a very short period of time and probably during the warranty period. This is the worst possible outcome! This means that screening without good coverage and good detection will most likely lead to increased field failures, probably immediately after delivery. Screening without appropriate monitoring is one of the major mistakes made by untrained engineers and is found in many published papers from the past and even today.

3.8 SELECTION OF DETECTION SCREENS

Detection screens are usually less stressful than precipitation screens and are aimed at making the patent defects detectable. The detection level is shown in Figure 3.4. Again, the level used to start the process of HASS optimization is an engineering estimate based on experience and on the data gathered during the HALT. The end result is insensitive to the initial estimate.

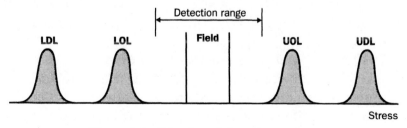

Figure 3.4 Selection of detection screens

A detection screen on a computer system (which takes a long time to test) might be a very slow temperature cycle (perhaps at 1 °C/min) between the lower and upper operational limits while running low-level, all-axis vibration and simultaneously running diagnostics. This would probably be a very poor precipitation screen but could be a very good detection screen.

An example of results of running detection screens is in order and is contained in the paragraphs below.

In attempts to set up detection screens in 1979, the author learned about the value of using all-axis impact random as an aid in detecting already precipitated flaws. In that case, one of the defects that led to the learning experience was due to assembly technique and the symptom was cracked solder joints where a ring connector penetrated a circuit board. After the solder joints were made, an assembly operation disturbed the joint to the point of cracking it. The thirteen joints were daisy-chained in a special test set-up so that an intermittent in any one joint would show up as an open circuit. A special tester was made up to read out the percentage of the time that the circuit was intermittent. Normal temperature cycling combined with single-axis random vibration on a special fixture which allowed the vibration to be aligned in any direction was utilized. An exhaustive test was performed to try to find the temperature and vibration conditions which would expose the intermittents. This test took the better part of a week without success no matter how the vibration was vectored, what was the overall level of vibration or the temperature at which it was applied. The (then) new all-axis system, now totally obsolete, became available for one afternoon and was immediately put to use. Thermal cycling while running high-level (for that time) vibration was begun. Nothing showed up. The vibration was then reduced to just a "tickle", below 3 GRMS, and the whole temperature range was tried. When the lower end of the range was attained, near −60 °C, the special test meter read 97% open dutycycle! Once this phenomenon had been discovered, low temperature and tickle vibration were used as a normal detection screen and for troubleshooting with outstanding success. It is completely amazing how effective the combination of low temperature and the tickle vibration worked out to be, and, as a result, it became a standard test method for the next 15 years. The major reason for the differences in detection are explained in [3], which describes the behavior of an early six-axis shaker.

At that time, the all-axis impact random vibration system became a necessary tool for troubleshooting systems which had failed in the system level screens since it was far superior for making the intermittents show up. This fact was rather grudgingly accepted by the rework personnel as the all-axis random vibration system was only available from 12 p.m. to 8 a.m. since production testing was in progress 16 hours per day except for most weekends or during line shut-downs. The troubleshooting could not wait until the weekends since corrective action should not be accomplished a week after the fact of producing defective systems because all systems made in the meantime could (and probably would) have the same defects. Therefore, troubleshooting was done from 12 p.m. to 8 a.m.

Later in time, in about 1988, the author learned by chance in a workshop associated with the advanced seminar at Harley Davidson Motorcycles that the tickle vibration worked even better if the level were modulated, between 1 and 4 GRMS, for example. Another successful technique was now available. In fact, the author designed a vibration system to modulate automatically over a significant range, ±20% from the set point in order to take advantage of this newly conceived technique.

Still later, in 1996, the idea of modulating the vibration from a full precipitation level to just as low as the six-axis shaker system could go was tried. Eureka! This worked even better.

In 1995, the idea of using a detection screen during HALT was conceived and tried. It was discovered that a substantial number of defects must have been missed in previous HALTs since the technique is extremely powerful in exposing failed joints of all types which will not show up under steady vibration levels, even under the steady all-axis vibration with temperature slewing. It takes the combination of slow temperature slewing combined with rapid vibration amplitude variation to generate a situation where the intermittents will be detectable. The author calls this the "magic combination".

90 HASS: HIGHLY ACCELERATED STRESS SCREENS

3.9 MODULATED EXCITATIONS

In order to evaluate a new idea of complete modulation of the vibration from the upper operational limit to the lowest level at which the shakers would operate, tests were performed during the advanced seminar workshop. In this workshop, a stereo amplifier was used as the unit under test. Great success followed as it was found that the complete modulation proved to be even more powerful than had the tickle vibration. Then, in 1996, the completely modulated vibration combined with temperature slewing was taught in both the introductory and advanced seminars as detection methods during a detection screen or during HALT step stress testing. Advanced users of HALT now run fully modulated excitation at appropriate times during the HALT step stress testing. Some of the defect types exposed by the modulated all-axis vibration and simultaneous thermal cycling are cracked solder joints, cracked component bodies, broken

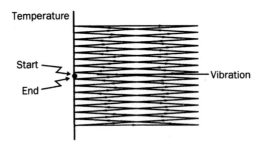

Figure 3.5 Modulated excitation in two dimensional space

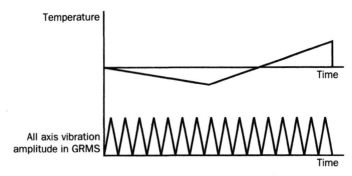

Figure 3.6 Modulated excitation in the time domain

component leads, cracked traces on PCBs, loose connectors and broken leads internal to components. Other types are likely, of course, to be found.

Modulated excitations are shown in Figures 3.5 and 3.6 in two different representations. In Figure 3.5, time progresses as the temperature and vibration path in the two-dimensional space is followed. This "search pattern" makes a fine grid of the two-dimensional space leaving little room for a detectable defect to hide. Figure 3.6 represents the same thing in the amplitude versus time domain. Both represent exactly the same thing, a fine grid search pattern over the temperature/vibration two-dimensional space.

Some results of the new combination of stresses used to find precipitated defects may help to illustrate the idea and the power of it. In the advanced seminar on HALT and HASS, an inexpensive stereo amplifier is used as the product to be run through HALT and HASS. During the step stress portion of HALT and during the detection portion of HASS, the conditions in Table 3.1 under which the precipitated defects were observed were noted (all on different units used in April 1996).

The RMS level relates to the set level. A system with automatic modulation of the vibration was used, so the variation on the set point was approximately ± 20% of the set point. A vibration change beyond the ± 20% would result in no detectable failure. A

Table 3.1 Defect observation conditions

Unit number	Temperature °C	Vibration level GRMS
1	−60	8
2	−38	28
3	−23	13
4	−15	25
5	−35	25–40
6	−45	33
7	−25	18
8	−40	1–3
9	−25	15

change of only a few degrees centigrade would result in no detectable failure.

None of the units used in the workshops in April had a hard failure! Every single one was intermittent and was stress-level dependent. In every one of the cases cited above, no detection was possible under any other conditions of temperature and vibration! Every unit was fully functional at ambient without vibration after the defect had been detected under the conditions cited above. This clearly demonstrates that not only are temperature and vibration together very synergistic in detection screens, but that the correct combination for each situation is different depending on the flaw type, location and degree of precipitation. It also shows, without a doubt, that detection must be done under stimulation as most (97%) of the failed units tested in the workshops in the last few years have passed all tests when stimulation was terminated. This is a normal situation and seems to be generally true for most electronic units. The same was found in years past when inexpensive two-way radios, the Radio Shack Space Command, were used in the workshops. The author's consulting experiences on many diverse products generally match the conclusions as well. It somewhat rare for a hard failure to occur and so one should think in terms of detecting soft failures in order to have successful screens. It is believed that many screening programs have been considered a failure because the failed items were not detected in the manufacturing plant because of the lack of a detection screen, but were found by the initial user, frequently upon first use. As an example, consider the environment in an airplane, bus or in an automobile – slow thermal cycling and varying all-axis vibration levels, just perfect for detection! It is no wonder that many defects are found by the end user instead of by the manufacturer when poor or no detection screens have been used.

An attempt was made in the workshop to drive some of the failures to a hard failure. Extreme temperature and vibration levels were used to further precipitation and the conditions at which the flaws were detectable were found in each unit. Only one of the units exhibited any change in the conditions under which the flaw was detectable. That one unit did not result in a hard failure, but the "magic" combination moved to a higher temperature and a higher vibration level by 15°C and 10 GRMS, respectively. This implies that many failures cannot be driven to a hard failure in any reasonable amount of time, perhaps ever. Continued attempts to drive an intermittent to a hard failure almost always results in a failure somewhere else than at the originally detected site.

A large computer manufacturer has reported that when a slightly modulated all-axis impact vibration was added to the combined screen, which had previously had a fixed commanded nominal level, then the fallout from the screen doubled immediately. This means that the detection had improved by a factor of two as the precipitation would not usually be increased by the lower vibration levels that were added.

Four points have been made regarding the screening of electronic equipment and are worth reiterating:

1. Most failures are not hard failures and are temperature and vibration dependent.
2. Modulated excitation optimizes the probability of finding the combination of stresses which provide a detectable state.
3. Monitoring *during* stimulation is absolutely essential.
4. Soft failures generally cannot be driven to hard failures.

3.10 DISCRIMINATORS

A discriminator is a test that allows the detection of a non-standard behavioral pattern. The discriminator may be some property or combination of properties of the device under test. These properties are not limited to those which are normally used or measured in a product, but can be anything that allows one to detect abnormal behavior and sort the defective ones from the good ones. Discriminators are always product unique and require knowledge of the product's normal behavior. If several parameters are used to form the discriminator, each parameter may be within normal limits, but the discriminator may show a deviation from normal. This is the power of discriminators. Suppose we had three parameters, *A*, *B* and *C*. A discriminator might be formed by taking *A*, *B* and *C* for each unit and calculating the ratio $A \times B^2 / C^3$. If this combined parameter could allow the investigator to sort out the units that are, or will become, unacceptable using this discriminator, then the discriminator is a good one.

One discriminator frequently found to be of use on electronic equipment is a microphonic noise signature taken during vibration. If the vibration-induced microphonic noise in a system can be characterized for a normal unit, perhaps by observing the spectral density of the noise, then it is sometimes possible to sort defective devices from good ones even though the defective ones have not yet failed to meet all requirements when exposed to normal conditions.

The microphonic signature technique was used in 1979 to sort out electro-optical devices which would later fail due to a certain defect type, but worked correctly at the time of the signature analysis. The characteristic was discovered by chance, but was successfully used thereafter without error. This is an example where the units "met spec", but were destined to fail later. The manufacturing people had to be shown repeatedly that a valid discriminator had been discovered that was like a look into the future in that it would predict, without error, those units which would later fail. The manufacturing people were initially insistent that if it "met spec", it was OK, and that the unit should be shipped as is. (This is an example of the "you took it over spec" syndrome.) Only after showing them that the units sorted out by the microphonic signature discriminator did indeed fail very early in a life test, did they finally accept the discriminator. There were two problems with acceptance of the test result to be overcome in this case. First, it could not be explained why the discriminator worked, which is really not relevant because it had been demonstrated that it did, and secondly, there was no spec on microphonic noise at all and the product's normal operating environment included vibration at a much lower level, 1 GRMS, than that at which the noise was measured, 20 GRMS. The discriminator was eventually accepted as a valid indicator because it *never* failed to sort out units with the particular flaw present.

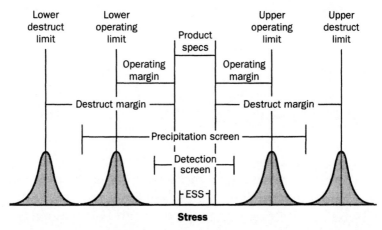

Figure 3.7 Precipitation and detection as selected from the HALT data.

3.11 SETTING UP THE HASS REGIMEN

In order to set up the HASS profiles, reference is once again made to Figures 3.3 and 3.4 which are combined as Figure 3.7. It is desirable to select stress levels for the precipitation screen beyond the operational levels (if possible). The short-term destruct levels must not be exceeded. Note that there will be little information on the distributions shown; it is just known that there are many distributions present with a different one for each failure mode. The distribution for the weakest failure modes are illustrated. The first estimate of a HASS profile is made based on what little is known of Figure 3.7. An estimate (guess) of what a good detection screen would be is also made. This detection screen must be within the operational regions for all stresses or else the product will not function properly. After estimating (guessing) what all stress parameters should be, including vibration, voltage, temperature, frequency of the power, humidity, contaminants, acidity of a fluid and so on, Safety of HASS is performed. This consists of cycling one product through the candidate screen 20 times. There is more discussion on this below. After passing Safety of HASS, production screening can begin with the knowledge that the HASS does not rapidly take life out of the product. Chapter 4 addresses this in some depth. During production, HASS tuning takes place.

In that methodology, the screen is tuned for maximum efficiency and minimum cost. Tuning is covered in Chapter 5. Safety of HASS is also called safety of screen (SOS). This is a good acronym to remember.

3.12 SAFETY OF HASS

It is necessary to demonstrate that there is sufficient life left in the unit after one or more HASS profiles have been performed upon the unit. Most units will ship after one HASS, but some units may have to be repaired and then run through HASS again to ensure that the problem(s) discovered in HASS have been solved. Occasionally, many cycles are performed before all problems are successfully solved and the unit clears HASS.

A method of demonstrating that HASS leaves enough life in a unit for field service has been selected and it requires that the unit survive 20 HASS profiles and still be able to pass all Qual tests. When the HASS profile is repeated in Safety of HASS, there is no rational reason for the number 20, it just seems to work well. One could use 10 or 30 and it would still work well. What is to be proved is that the profile picked is not destructive to good hardware. If the hardware can pass 20 HASS profiles, then it is concluded that not more than 5% of the life is taken out by each HASS profile. Therefore, running a few HASS profiles in production on the product should leave plenty of life in the product. How can it be proved that enough life is left in the product after a few HASS profiles? One can then run the same product through a complete Qual test series and, if all are passed, can conclude that the product still has adequate life remaining after HASS. If the product cannot pass 20 profiles without reaching end-of-life types of failures, then either the stresses are too high or the product is too weak. Both situations are addressed in Chapter 5. A novice will usually pick stresses that are too low and iteration in HASS tuning will result in increases that do indeed expose the flaws in the product in only one HASS profile. If the product is too weak, this too will come out in HASS tuning. Both are covered in Chapter 5. These are mentioned here to give the reader some comfort that there is a rational approach to prove that what is suggested will indeed work in the end.

A question never to ask is, "How many profiles can the product withstand?" This sounds like a good question, but obtaining the answer may take years and cost millions. For example, an associate was asked this question by his manager and he set out to find the answer [4]. He set up 10 products in the chamber and proceeded to run 1,251 profiles complete with proper detection screens. At this point, he called the author and asked how to get out of this seemingly endless test. It was suggested that he get the program manager to agree to end the tests if all products could still pass full Qual tests. The manager agreed and all units did pass in every regard. This clearly answered the *correct* question which is, "Do the products have enough life left after a few HASS profiles to survive a normal lifetime in the field?"

In another situation, circuit boards used only during HASS to fill up a system temporarily with all options so that the complete bank of tests could be run successfully, completed over 4,000 HASS profiles without a single failure in any of the 13 boards. In still another example [5], 100 repeats of the HASS profile were performed with no end-of-life failures. It seems that thousands of repeats can be performed without end-of-life failures if the proper HALT technique has been followed.

It becomes apparent that the question, "How many profiles can the product withstand?" and its corollary, "How much life does one HASS profile remove?" are not questions which can be reasonably answered. The author is satisfied if 20 profiles can be run with no end-of-life failures occurring. This satisfaction is based on experience and not on science. If one knows what the weakest link or links in the product are, then one can, in principle at least, use the physics of failure approach to answer the questions if and only if one also knows exactly what the field environments are on a time history basis, which is usually not known with any accuracy. In the end, trust is placed on the information and intuition built up by many HALT and HASS programs that have been properly performed.

3.13 THE EFFECTIVENESS OF HASS

Some users of screening have attempted to prove that the screens are effective by building seeded samples (units with intentional flaws) and proving that their screens can indeed expose the flaws. This currently is thought to be an exercise in futility in the author's opinion, although the subject used to be taught in the author's seminars. What has been learned over the years is that many flaws can be built into the product and no amount of screening will expose the flaws because the stress at the flaw is not sufficient to cause a failure, even with a gross defect. A similar flaw in another location may be exposed almost immediately if the stress there is high enough. If one knows enough about the stress distribution in the product and places the flaw in a critical location, seeding samples can work. A problem is that many of the users of HALT and HASS are not trained stress analysts and cannot seem to place flaws in the "correct" location. This leads to many questions from the program managers who then start to request proof of various things and the whole exercise becomes one with no end. It is suggested that one not attempt to prove the effectiveness of HASS by seeding flaws for these reasons.

A seeded sample exercise is performed in the advanced HALT and HASS seminar given by the author. Very seldom does an attendee succeed in seeding a sample which will work when put into the stress chamber and then fail in the first HASS. This demonstrates the difficulty of performing seeded samples. Continued stressing at ever higher levels almost always results in a failure other than at the seeded defect.

When using true HASS in production for the first time, one will find more flaws than ever found before and will probably be accused of breaking good hardware and will hear all of the usual misconceptions expressed such as, "Of course it broke, you took it over spec!" The real problem to be faced is not that of effectiveness, but to prove that the screen does not damage good hardware and Safety of HASS does this.

Sometimes attempts are made to tune a screen by increasing the stresses until seeded samples fail. This can lead to failures at low stress locations, where the seeds may have been placed. This would mean that severe overstress can exist at the high stress points leading to end-of-life failures of good hardware. This is the basic flaw in the approach of increasing the stresses until seeded samples fail. At any rate, Safety of HASS should *always* be run no matter how the screens are selected. Omission of this critical step has led to many field failures on many different types of hardware. This blunder is usually made by those who do not take the time and spend the money to obtain complete training in the techniques which are really rather simple when carried out accurately and completely.

3.14 FAULT COVERAGE AND RESOLUTION

In addition to running the best detection screens from a stimulation point of view and getting the flaw into a detectable state, it is very necessary to have excellent fault coverage. It has been found that most companies think that they have high, perhaps 95%, fault coverage, when indeed they really have much less, sometimes drastically less. The actual test coverage can be accurately assessed using automated fault injection (AFI) techniques which are discussed in Chapter 8. The subject of resolution; that is, the ability to discern *where* the fault is, is also covered in Chapter 8 on Software HALT.

3.15 HASS ON FIELD RETURNS

Detection screens should be used on equipment returned from the field as defective, since it is assumed that a patent defect is present. It is noted in passing that it is recognized that non-defectives are frequently returned from the field for various reasons caused usually by the press of time to "get it running A.s.a.p!" Field repair people are inclined to pull whole sets of boards or boxes and replace them, when maybe only one of the set truly has a problem. A precipitation screen may not be necessary on field returns as the patent defect(s) present may be exposed by a much more gentle detection screen. If the detection screens do not suffice, then a precipitation screen followed by a detection screen would be in order. In the case of field returns, it may be prudent to simulate the field conditions under which the failure occurred, if these can be ascertained. These conditions might include atmospheric pressure, temperature, vibration, voltage, frequency, humidity and any other relevant conditions. The military, airlines, auto manufacturers and others, too, would be well advised to follow this course of action as No Defects Found (NDFs) account for more than 50% of field returns as an industry-wide norm. The author has seen several cases where NDFs were at a level of 100%. In these specific cases, modulated excitation exposed 100% of the problems!

3.16 CONCLUSIONS

This chapter has discussed the initial selection of stress regimens using the immense accelerations (time compression) caused by increased stresses. The accelerations are used to advantage in HASS to achieve the shortest effective period that will precipitate flaws from latent to patent and then achieve a detectable state of the flaws. HASS optimization techniques can be applied in order to determine the optimum duration of the screen and this is covered in Chapter 5. Modulated excitation has been discussed and recommended as a way to obtain a detectable state.

Note that most of the benefits of HASS can only be attained if a proper HALT has been performed, including corrective action taken for all opportunities for improvement. The same comment applies to "standard" techniques because fatigue damage is imparted to all of the hardware no matter how the screening is done. If small margins exist between the field environments and the destruct limit, the product will not be effectively screenable by *any* technique which uses stresses. In that case, only inspections or other non-stress methods could be used. It is noted in passing that even power-up-and-test uses up life since electromigration is progressing during this time and electromigration is a wearout mode of most solid state devices and in some assemblies such as power supplies. The heating and cooling associated with power-up-and-down also leads to fatigue damage, although perhaps on a very minute scale. The examples given also demonstrate the concomitant equipment, labor and facilities cost savings of a properly done HALT and HASS sequence.

"Tickle" vibration has developed into a much more sophisticated and effective fully modulated all-axis broad-band random vibration combined with thermal cycling at a very slow rate to obtain the "magic combination" of stimulation for detection. This methodology amounts to a search pattern. Good test coverage is essential for good detection.

One might express the success of HASS in the following equation.

$$P(S) = P(P) \times P(D) \times P(FA) \times P(CA), \qquad (3.1)$$

where

$P(S)$ is the probability of success

$P(P)$ is the probability of precipitation

$P(D)$ is the probability of detection which consists of the product of two terms which are the probability of putting the flaw into a detectable state and the test coverage, i.e. $P(D)=P(DS) \times P(COV)$

$P(FA)$ is the probability that failure analysis will be adequately done

$P(CA)$ is the probability that corrective action will be adequately done.

The entire relationship is multiplicative, so each part of the discovery and fix process must be present in order for the methods to work successfully. If any one part is missing, then little progress can be expected.

Once again, it is cautioned that many programs reported in the open literature include many of the critical mistakes that are delineated in Chapter 10, and so those reports should be viewed critically before using any conclusions drawn therein.

REFERENCES

1. Caruso, H., Significant subtleties of stress screening, *Proceedings of the 1983 Reliability and Maintainability Symposium*, pp. 154–158.
2. *Quality and Reliability Engineering International*, **6**, (4), September–October 1990. The entire issue is relevent.
3. Hobbs, G. K., Six degrees of freedom vibration stress screening, *Journal of the IES*, November–December (1984).
4. McLean, H., Highly accelerated stressing of products with very low failure rates, *IES/ESSEH Workshops on Accelerated Stress Applications*, Vancouver, WA, 17–19 March, 1992.
5. Hopf, A. M., Highly accelerated life test for design and process improvement, *Proceedings of the IES*, (1993), pp. 147–155.

CHAPTER 4
Proof of Hass

4.1 INTRODUCTION

The concept of Proof of HASS has been in use since 1979 when it was called Proof of Screen, and the technique has been improved several times since then. The basic idea has two parts:

1. demonstrate that the chosen screen leaves the products which have been through the screen with sufficient life left in them to survive a normal lifetime of field use (Safety of HASS), and
2. to demonstrate that the chosen screen does indeed find the latent defects in the product (proof of effectiveness).

Safety of HASS is absolutely essential before production screening begins. Effectiveness becomes a given if other parameters are met. Both of these will be discussed in turn.

There is no guarantee, a priori, that a screen will precipitate and detect any or all latent defects present in a product. This must be proven by some technique or experiments as part of the selection and tuning of the screens. It is assumed in this chapter that test coverage; that is, the ability to detect a patent defect via a functional test, is very good, in the range of 90%–95%. This assumption is generally *not* true and will be covered in Chapter 8. It cannot be overemphasized how important good coverage is in the testing of electronic systems. If something that is wrong cannot be observed, then nothing can be improved. In that case, only catastrophic hard failures, perhaps 1%–2% of the total, would be found and could be fixed. Equivalent statements could be made about mechanical systems. It is generally true that if it cannot be observed, then nothing can be done to improve it.

It is emphasized that every cycle of stress to which a product is exposed does remove some life from the product; that is, does fatigue damage. Fatigue damage as used herein means reduction in remaining life. Even powering up and monitoring a product takes life out of it, even if only a very minute amount and quite beyond measurement. It is therefore crucial to understand the wearout and other failure modes of the product when developing HASS profiles. One must be assured that the product is shipped with enough life left in it to satisfy the requirements of the end user and to have some margin left beyond that to cover unexpected events and variations from unit to unit. That is, the entire population should provide an appropriate service life before wearout. Safety of HASS will provide assurance that the two conditions mentioned above are fulfilled.

4.2 SAFETY OF HASS

One obvious way to demonstrate that it is safe to ship the products after one or maybe a few HASS regimens would be to take a few of these units and place them into the field environments and simply run them until they failed. If a good job has been done in HALT and if the assembly processes are reasonably good, then nothing may be found for many years. By such time, the product would be obsolete and any answer obtained would be meaningless other than to prove that the job had been done correctly or incorrectly. Our feedback loop time would be many years. Clearly, this approach, although very thorough and exact, would not satisfy our requirements of releasing a product to full-scale production with some knowledge of what the result of HASS has been, particularly in regard to remaining life.

Another approach is to run a few HASS profiles on a product or two and then assess how much life is left by some accelerated test or analysis scheme. This would work if the field stress exposure levels were known, which is generally not the case, and if it were known which wearout modes would be the critical one or few, which is also not generally the case. This leaves a conundrum in place: accurate tests take too long and accurate analyses using the physics of failure approach is practically impossible due to the plethora of possible failure modes and sites, which generate a nearly infinite set of necessary analyses. Therefore, one must try to obtain some estimate of life left or get some estimate of the probability of success of the product to withstand a lifetime of field use. A way to do this is to run a series of HASS regimens on a product and try to assess if there is enough life left for a normal use pattern of the product after the series of regimens. This approach will not give an exact answer, but it will provide the practical answer that enough life is left after a few HASS cycles.

The series of HASS profiles has been arbitrarily selected as 20. If the product fails in a wearout mode just at the end of the last regimen, then it can be noted that one profile took out 5% of the product's life. From this, one can conclude that after one HASS profile at least 95% of the life is left, at least on this one-test item. It must be borne in mind that there is a distribution over the entire population for many different failure modes and usually only a few are tested. This is part of the reason for performing 20 HASS cycles instead of just a few.

But, the question remains: "Is this enough?" One can then run the same product through a complete Qual test series, and if all tests are passed, one can conclude that the product has plenty of life left after HASS. If the product cannot pass 20 profiles, then either the stresses are too high or the product is too weak. Usually, a novice will pick stresses that are too low and iteration in HASS optimization will result in increases that do indeed expose the flaws in the product in only one HASS profile. If the product is too weak, this will come out in HASS tuning. It usually occurs, if a proper job of HALT has been accomplished, that products can pass a few HASS regimens and still pass full Qual tests. This proves that it is safe to use HASS on production hardware and then ship the units to the customer. A logic diagram is shown in Figure 4.1.

A very natural question which arises is, "How much life is removed by the HASS?" One never knows the answer to this question and it is suggested that one never ask this question since to

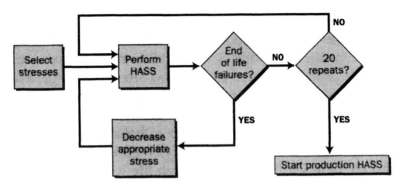

Figure 4.1 Safety of HASS

obtain the answer can be very expensive (as mentioned earlier). The author has observed one case where this was asked and, in order to attempt to answer it, 10 units were run through 1,251 sequential HASS cycles. There were no failures during this extended and surely very expensive test series. Finally, management was convinced to allow the termination of the test if the units under test could still pass Qual tests. Every one passed! One still does not know how much life was taken out by the HASS; however, one does know that there was still plenty of life left after the HASS. Even after 1251 HASS cycles the product could still pass a Qual test, so it must still have been OK for field service. The relevant question to ask is "Is there enough life left in the product after the HASS?" Repeated HASS; i.e. 10 or 20 cycles and then a Qual test answers this question as economically as is possible. The author prefers not to do any stresses in the Qual test that have already been done in the Safety of HASS series since it seems to be extreme over-testing for no useful reason. Also, the HASS levels are usually far higher than the Qual levels, especially in terms of fatigue damage accumulation, so doing more fatigue damage does not seem to make sense.

An additional matter needs to be addressed. How does one know that the HASS has not set up some long-term failure mode which has not shown up yet? In order to answer this question, 10 or 20 cycles of HASS are performed then finally a Rel-Demo is performed on them to see whether any long-term failure modes have been excited. On electronics, a failure mode that might exist is that of cracked component bodies, which may not show up for a significant time. This may be accelerated by using Highly Accelerated Stress Tests (HASTs), which is high pressure, high temperature and high humidity all applied simultaneously. It is suggested that all HALT and Safety of HASS units be subjected to a HAST in order to determine whether any undiscovered long-term failures modes are present.

As time progresses, the processes with which the product and its components and subassemblies are built will change (even with a frozen design). In order to address this, one should repeat HALT in order to ensure that the capability distributions have not degraded. If they have degraded; that is, decreased, one should attempt to gain the margins back. *Degraded margins will invalidate the Safety of HASS.* This can be a major problem since the HASS profiles would have been based on the margins as originally extant, not on the degraded margins. HASS could then take the products too close to end of life or even past end of life. This situation has been observed in companies where only part of the techniques presented herein have been followed. Outsourcing can contain this as a potential hazard if the outsource does not do all six steps of the reliability program: precipitation, detection, failure analysis, corrective action, correction action verification and documentation of knowledge gained.

4.3 PROOF OF THE EFFECTIVENESS OF HASS

How to prove effectiveness will be discussed even though it is thought not to be necessary. In order for any screen to be effective, it must precipitate and then detect any latent defects which would be exposed by the normal use environments, including storage and shipping.

When using HASS with appropriate tests to detect precipitated defects, it is seldom necessary to prove that the screens are effective in a separate experiment because:

1. production fallout from the screens will probably be higher than ever before and so this fact alone nearly proves that the screens are effective

2. HASS optimization in Chapter 5 will show that additional fallout is not occurring after one HASS cycle after optimization has been accomplished

However, field failures are the only real, easy way to do proof of effectiveness and this is done automatically by product exposure to field environments and use. It may happen that additional fallout occurs in the field, and, in that case, some critical stress or test has been missed and corrective action would be appropriate. It is assumed, and is indeed essential, that the product is effectively monitored with good coverage during the stressing or precipitated defects will be missed; i.e. not observed. Precipitation without detection will generate a complete catastrophe since the customer will probably quickly observe the faults missed in detection. There is nothing more aggravating to a customer than to experience out-of-the-box failures. This is exactly what will happen if good detection screens with high coverage are not in place during the stimulation. HASS without high coverage monitoring is less than worthless. This mistake is one of the most frequently occurring mistakes observed in the open literature and is completely unnecessary.

If one insists on evaluating the effectiveness of a HASS regimen, it is indeed possible but is fraught with difficulty and frustration. The method used is called the seeded sample method, wherein intentional defects are placed in the product and then the HASS is run to determine whether the screen can precipitate and then detect the latent defect. Such a test is outlined in Figure 4.2. Proof of effectiveness is currently thought to be an exercise in futility in the author's opinion, although the subject was taught in the author's seminars in the 1980s. What has been learned over the years is that many flaws can be built into the product and no amount of screening will expose the flaws because the stress at the flaw is not sufficient to cause a failure. A similar flaw in another location may be exposed almost immediately because the stress there is much higher. If one knows enough about the stress distribution in the product and places the flaw in a critical location, seeding samples can work. A problem is that many of the users of HALT and HASS are not trained stress analysts and cannot seem

PROOF OF THE EFFECTIVENESS OF HASS

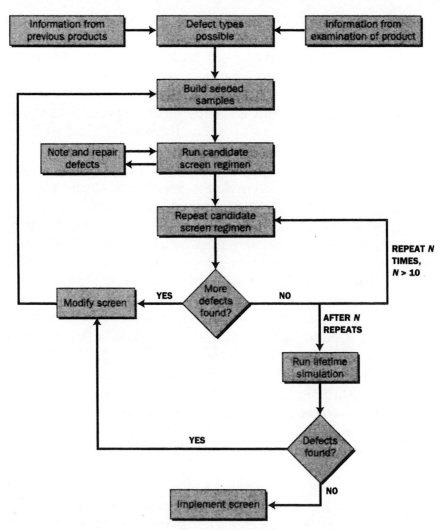

Figure 4.2 Seeded samples to prove screen effectiveness

to place flaws in the "correct" location to obtain a failure. This has been demonstrated many times in the workshop portion of the advanced seminar where the attendees attempt to perform the experiment, usually without success. Performing the seeded sample experiment in a production program usually leads to many questions from the program managers who then start to request proof of various capabilities and the whole exercise becomes almost endless. The author has observed several companies which expended six months or more trying to perform a successful seeded sample experiment. One company just gave up on the idea of HASS because they could not build seeded samples which would fail in HASS.

Seeded samples will be discussed so that the reader can gain an appreciation of the difficulty and expense of the task if undertaken. As in all screening, one must continually exercise good engineering judgment as one proceeds. One is not performing a MIL-SPEC type of test, but is trying by whatever means necessary to precipitate and then detect latent defects which would cause failures in a normal use situation, including shipping and storage. One begins by listing all the defect types which could exist in the product by drawing on experience and by observing the product piece by piece and listing all the possible defects which could occur in the components, in the subassemblies or in the processes. Previous similar products will aid in the listing. Several products containing samples of each of the defect types in various locations are then built. Note that the number of products built for the tests may become quite large and this is one of the reasons for not doing this experiment, as well as the fact that building samples which will fail is a difficult task indeed. Note that some pieces of hardware can be used repeatedly in the experiment since correctly built hardware can usually be exposed to HASS for hundreds or even thousands of cycles without failure.

Just as an example, consider the ubiquitous defective solder joint. A unit with good, marginal and poor solder joints in various locations is fabricated. Note that screens are location-dependent and the stress levels to which the joints are exposed is not constant and a poor joint may survive in one place and a marginal joint may fail in another. If the detection portion of the screen cannot detect a defective solder joint; that is, an open or an intermittent, at the particular location in question, then the defect will not be detected even if precipitated. These are two reasons why running proof of effectiveness this way is a very difficult thing to do.

The seeded sample, a sample with intentionally defective solder joints, is now subjected to the selected HASS sequence (Figure 4.2). Particular stresses used are all-axis broad-band random vibration, rapid thermal cycling over a broad range, power cycling and full monitoring, at least during the detection screen portion of the HASS. Other stresses are used if appropriate for the product and the flaw types expected. All defects detected during the first running of the HASS are noted and repaired. The HASS profile is now run a second time to see whether any more defects are noted. If so, then a HASS change is made since the selected HASS did not find all the defects in one profile. If optimization has already been done, this should not occur, and if optimization has not been done, then it will be accomplished as the test proceeds. If nothing is found in the second HASS profile, then a third is performed and so on, following the flow of Figure 4.2. If ten or more HASS profiles can be run with no more defects being found, then the profile has probably precipitated and detected all that it is going to. Finally, a lifetime simulation is performed to determine whether the field environments and other conditions will expose something that the repeated HASS profiles did not. If the product(s) all successfully pass the life test, then it is proven that the HASS profile exposed and detected everything that would show up in a normal lifetime. This means that the HASS is perfect (at least this test says it is). Other types of seeded sample defects will also have to be run to verify that the HASS profile finds everything. Note that there is an almost endless set of flaws and locations which could be explored.

As can be seen from the above discussion and by studying the flow chart of Figure 4.2, screen optimization will occur during the seeded sample tests if it has not already been done. Generally, if seeded samples are done at all, they are performed before production screening so the screen optimization will not yet have been performed. This will make the seeded sample tests very long and complicated. This is another reason that the author suggests not performing seeded samples.

There is another problem that crops up. Suppose that some defects are not exposed by the HASS regimen. Some observers may say that the screen is not good because it did not expose an obvious flaw. Now one must argue that the stress at the particular location of the flaw was not sufficient to cause a failure and so the fact that HASS did not expose the flaw is not important. This argument is frequently met with disbelief and so further tests must be run to confirm the point. More time and money are then spent trying to confirm that the defects as seeded will not cause a field problem. Again, it is recommended that seeded samples not be performed because of the time, anguish and money involved. In spite of this recommendation, the author has seen several companies get into the seeded sample tests and become so involved that the HASS program was delayed by six months in one case and by eight months in another. This delay set back the benefits of HASS by that much and also consumed substantial sums of money in the process. It could even occur that HASS is never implemented because of questions brought up and not answered (or even being unanswerable) in the process.

The running of seeded samples is fraught with frustration since often very obvious flaws will not be precipitated and will not fail in the Qual Test either. Be prepared for this outcome since it will most likely occur and is somewhat frustrating to those first attempting the method, and to their management as well. It is hoped that the difficulties in the seeded sample method have been effectively pointed out so that this approach will not be tried.

4.4 SUMMARY

The concepts of Safety of HASS and HASS effectiveness have been introduced and discussed. The concept of Safety of HASS has been found to be extremely valuable to the point of being mandatory in the many years of application in the military and commercial fields. It is the author's opinion that no production screens should be used without first performing Safety of HASS to demonstrate that HASS is not destructive to good hardware and that it leaves sufficient life in the product for normal field use. HASS effectiveness is excellent in concept, but is extremely difficult to perform successfully in practice. It is suggested that it be omitted unless absolutely demanded by management. Screen optimization, covered in Chapter 5, will demonstrate all that needs to be done in the way of HASS effectiveness.

CHAPTER 5
Hass Optimization

5.1 INTRODUCTION

HASS is very much worthwhile when done in an optimal fashion. However, when done in a less than optimal fashion, it can become less cost effective. Even then, it may exhibit very large ROIs and so could still be worth doing. HASS optimization is a process which is intended to minimize the time required to carry out HASS, and, in the process, minimize the cost. Minimization of the vibration time, the number of thermal cycles, the number of power cycles, the diagnostic time and many other stresses or tests as well lead directly to a minimum cost HASS.

It happens in most screens that there is a definite limit as to how much HASS is necessary in order to precipitate and detect the latent defects in any product. This corresponds to reaching the bottom of the bathtub curve after having gone through the infant mortality stage of the curve and found most of the early life failures. Any additional screening after that point just uses up useful life and costs money to do so. The task to be performed is to identify the optimum duration of any given screen. This will inherently minimize the cost.

The optimization process is an experimental one and requires a large sample size if the as-built quality is high and there is little in the way of flaws to be found. The only economical way to do the optimization is to use production hardware for the process. This is satisfactory since all of the hardware subjected to the process can be shipped with the confidence that it is not worn out and is not flawed either if correct procedures have been followed.

5.2 THE HASS OPTIMIZATION PROCESS

The process begins by running Safety of HASS as discussed in Chapter 4. After it is shown that the hardware can be screened multiple times without experiencing end-of-life failures, then it is clearly safe to carry out HASS on the hardware four times and still ship it for revenue. In this case, it obviously has at least 16 more HASS cycles left in it. The real-world lifetime usually represents less than one HASS cycle of fatigue damage accumulation.

The production HASS is set up to repeat four times. Four is an arbitrary number picked by the author based upon intuition and experience. It will be seen that the arbitrary number could be three or five with little effect on the outcome. It is generally possible to carry out HASS on early production hardware repeatedly since there is usually more HASS capacity than there is production volume requiring it. Production usually increases from low volume to high over some short period of time due to many factors such as limited components or subassemblies or even the ability to test all the hardware adequately. Sales may also grow slowly due to market acceptance. For whatever reason, many production lines do not start up at the eventual 100% of capacity and that leaves the capacity needed to run the optimization experiment at the start of production, which is the correct time to perform it.

Data is taken on the failures during the repeated HASS cycles and is plotted versus the cycle number at which the defect is found. If more HASS cycles are needed beyond the first one, observation of the failure modes can give some indication as to what should be changed so that the defects would be precipitated and detected during the first HASS cycle. An example of such a diagram is shown in Figure 5.1.

The flaws have been broken down into those precipitated/detected by vibration, temperature and voltage. This is easier said than done, it is realized, but an example is needed for clarification. If the flaw seems to be a flaw that could be uncovered by vibration,

THE HASS OPTIMIZATION PROCESS 115

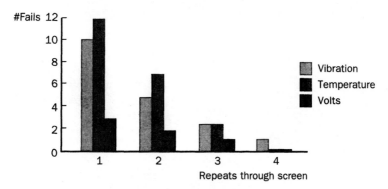

Figure 5.1 HASS optimization

then one would add slightly to the vibration level in an attempt to uncover the defect in less time, in fewer passes through HASS. It is desirable to find all the defects in only one HASS. Caution in increasing stresses is in order since the relationship between stress level and the number of cycles is usually exponential in nature as shown in Chapter 2 for one particular type of stress. Others are similar in nature. See Chapter 7 for more relationships.

Assume that failures are still showing up in the fourth HASS cycle and that the failures have been exposed by vibration. In order to get four times the fatigue damage in one HASS, one could either increase the vibration duration by a factor of 4 or increase the GRMS level slightly so that the stress raised to the 10th power is about four times as much. Doing the calculation results in increasing the GRMS level by a factor of only 1.15 assuming a beta of 10 for the slope of the S–N diagram. It is seen that very little increase in stress is required to accomplish the acceleration of precipitation; however, an increase of four times would be required in the duration of the vibration. Similar effects occur for other stresses. It is always suggested to try increasing the level first instead of the duration since the costs will be much less this way. Some nonlinearities may exist, particularly in vibration, where electronic equipment is typically very non-linear in the vibration input–output sense. In spite of this, it has been the author's experience that the iteration methods described herein have always converged to a satisfactory optimum quite quickly.

After adjusting all stresses to get the needed estimated fatigue

116 HASS OPTIMIZATION

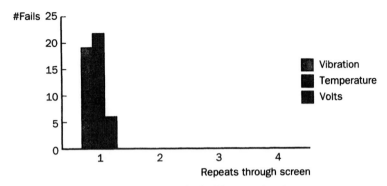

Figure 5.2 Final HASS optimization

damage accumulation rates based on knowledge of the physics of failure, Safety of HASS is again performed on production hardware and then the multiple screens until enough data is obtained to either confirm that all the defects are being found in only one HASS or that further tuning is necessary.

An example of what might occur if good estimations of the adjustment were made is shown in Figure 5.2. Now all defects are found in the first HASS for the sample tested and the HASS optimization is finished.

Note that minimizing the vibration duration, minimizing the number of thermal cycles and so on is the same as minimizing the total cost of equipment required (shakers, chambers, monitoring equipment) and also the consumables (electrical power and liquid nitrogen).

It may occur that all defects are found in the first HASS and the figure may look like Figure 5.2 on the first attempt. The question, then, is: "Is the HASS selected more than necessary to precipitate and detect the flaws present?" The way this case is handled is to reduce the HASS levels and/or durations in selecting a new HASS profile. In this case, it is suggested that the number of cycles or the duration of the stress be reduced as that is the most cost effective change to make. The optimization is then performed as stated above to see if the selected profile is sufficient. Figure 5.3 illustrates the logic of the sequence. It is seen that underestimating the optimum is better than to overestimate the optimum in the first estimate of a HASS profile as the overshoot case requires several,

THE HASS OPTIMIZATION PROCESS 117

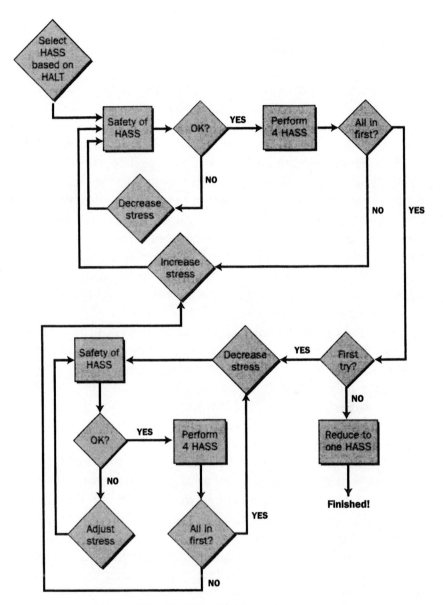

Figure 5.3 HASS optimization logic diagram

and perhaps even many, trials and the undershoot case will only require a few trials.

5.3 CONCLUSIONS ON HASS

HASS optimization is usually accomplished over a very short period, perhaps a few weeks. It must be done on production hardware in order to get a large enough sample size to ensure some reasonable accuracy in the results. All hardware exposed to multiple HASS cycles is generally shipped for revenue and is expected to be as good as any other units due to the fact that Safety of HASS has been run on the design before each optimization run. The optimization will result in minimum time and cost for the HASS sequence. This technique has been used to reduce substantially the costs associated with HASS and is highly recommended. It has been found that very substantial amounts of money can be saved by the optimization process and that it takes little time and money to perform.

During the optimization process, emphasis should be placed on minimizing the number of thermal cycles because this generates most of the expense in HASS. It has been found in many programs that one precipitation thermal cycle and one detection thermal cycle is sufficient. All-axis vibration, power cycling and other appropriate stresses are, of course, simultaneously applied with the thermal cycling.

If large design margins do not exist, then it may not be possible to do the optimization or even to screen effectively at all since when enough screening has been done to get the flaws out, not enough life will be left for field use. This is the usual situation if HALT has not been done properly or has not been done at all.

It has been observed in published papers that some untrained engineers attempt to tune a screen by increasing the stresses until seeded samples fail. This can lead to a major problem because some of the seeds which fail may be at low-stress locations. At the very high-stress locations where there may be no seed, extreme fatigue damage may occur, but, without failure, no clue is given as to the high-fatigue-damage conditions. A screen of this type would leave little, or maybe no, life in the product after HASS. As a result, many wearout field failures could occur in early field life.

Safety of HASS should always be performed to prove that a screen leaves enough life in the product for field use. Omission of this crucial step can be catastrophic to a company's financial well-being. Safety of HASS should be repeated from time to time or re-HALT can be performed instead to measure the various margins and ensure that they have not degraded. Degradation of margins from those present during Safety of HASS can lead to end-of-life failures during the warranty period. The author has observed this very situation a number of times when he has been called in as a consultant after failures have begun. The margins must be maintained in order to avoid this condition. Re-HALT on an ongoing basis is the best insurance against such an event, which is a certainty unless something is done to prevent it. The second law of thermodynamics on entropy says, "A system will always go to a lower organizational state unless something is done to prevent it." One can count on a production process to degrade unless something is done to upgrade it continuously.

After HASS optimization is completed and the program runs its course for some time, the number of defects found in the proven and tuned screen may decline to a level where no screening at all would be satisfactory except for the ever-present danger that something will change. History has shown that changes will happen, but one has no idea as to when a change will occur or what will change. In this situation, it may be optimum to shift to a sample plan and screen only a portion of the total production output as long as all processes are under control, the screen failure rate is acceptable and the field reliability is also acceptable.

5.4 HIGHLY ACCELERATED STRESS AUDIT (HASA)

HASA is simply HASS on a sample basis. Note that, when sampling, the units not in the sample will be shipped with no screen at all. This implies that one should be satisfied with the prescreened reliability since that is what will be achieved in the field for all products which are not in the (small) sample.

There are several ways in which to select a sample size. In the early 1980s, the author was involved in a program wherein the sample size was selected by starting at 100% until the results were satisfactory, i.e. the failure rate was low enough to satisfy the internal and external requirements. Then the sample was reduced to 50%, 25%, 12% and finally down to a minimum of 6%. If at any time the failure rate exceeded the selected maximum, the sample was increased to 100% until corrective action brought the failure rate back to an acceptable level. In this example, the sample size paced the production flow and the total production value was in the order of US$10,000,000 per hour at a 6% sample. At a 100% sample, the production value was only US$600,000 per hour for a loss of US$9,400,000 per hour! Therefore, management was extremely interested in fast corrective action, which they, of course, received. The *initial production* products exhibited a *measured field MTBF* of 500,000 hours. This result was outstanding in 1983 and is still quite good today.

Another approach in the selection of sample size is to use statistics to determine sample size and the associated risks. The basis for this approach can be found in any good text on statistics and so will not be covered here. For an excellent and concise account of the application of statistics to HASA, see [1].

Sampling makes a lot of sense since HASS is expensive, albeit worthwhile. As a company matures and becomes "product smart", little will be found in HALT and little will be found in HASS. That is the time to reduce the screening costs by advancing to HASA. Some advanced companies have successfully sampled at rates as low as 0.3%.

REFERENCE

1. McLean, H., Highly accelerated stressing of products with very low failure rates, *1992 Proceedings of the IES*, pp. 443–450.

CHAPTER 6
Uniformity and Repeatibility

6.1 INTRODUCTION

It is generally thought that uniformity and repeatability are essential to screening and other testing activities. It has been found by the author and others that neither tight uniformity nor tight repeatability of the vibration or thermal profile is necessary in a comprehensive HALT and HASS program. This statement is extremely dependent on the large margins obtained in HALT and is generally not true otherwise.

6.2 UNIFORMITY

A one-year-long experiment was run at Santa Barbara Research Center in 1980 on a program which was reported in [1]. It is extremely important to note that the product to be screened had been run through Design Ruggedization, a precursor to HALT, and had been severely ruggedized. The electrodynamic shakers and thermal shock systems available at that time could not damage flaw-free units even with week-long exposures to the maximum stresses that these equipments could generate. When production Enhanced ESS, a precursor to HASS, was begun, a new type of vibration and thermal system was used. In that program, an all-axis shaker and thermal chamber system was used for the

Enhanced ESS of an electro-optical device, a proximity fuse for a Sidewinder Missile. The fixture mounted on the shaker and within the thermal chamber held 12 target detectors. The thermal and vibration profiles and levels at each of the 12 screening positions were very much different. The thermal profile varied by 20 °C on the maximum and minimum side and the temperature ramp rates at some positions were twice that at other positions. The vibration level as measured by the GRMS level varied by +/− 50% and the vibration PSD varied by up to 60 dB from position to position. This variation was not expected nor thought to be acceptable since the classical paradigm of uniformity was in place in the author's head at that time.

Since the thermal cycle/vibration equipment had to be used regardless of the uniformity, a Safety of HASS (then called Screen Clearance, see [2]) was performed. The products were run through four screens sequentially with the vibration at maximum level, only 11 GRMS all-axis impact random, and with the thermal system running at the maximum possible ramp rate, which was 15 °C/min (chamber air). This repetition through the maximum possible screen did not result in any failures of any kind on any of the 12 units. Subsequently, some units were run through many of the intended production screens to determine whether damage to the unit under test was possible with this screening equipment. No matter what was done to the units, the screening system could not damage good units. This test gave some comfort that the screens were safe in terms of product damage due to the variations in the thermal and vibration environments. Production screening with the customer's approval was begun. As a matter of curiosity, screen-precipitated and detected failures at each of the 12 positions on the fixture were tracked for a complete year and then Pareto charts were drawn for each position. The charts turned out to be quite similar. This meant that each of the 12 very different screens was equally effective! Sufficiency of screen had been proven by performing multiple screens on production hardware and demonstrating that the bottom of the bathtub curve had been reached.

A graphical explanation will help to clarify the concept mentioned above. The target detectors had been ruggedized to the extent that they were extremely robust and could take much more than "spec" vibration and thermal cycling because it was felt that the "spec" was way too low for the intended application and that the road to good quality was large design margins and then good process control. The hypothesized failure rate before ruggedization is shown in Figure 6.1, representing the failures that would have occurred if a long simulation of the field environments had been run. The pre-ruggedization product had design- and process-induced weaknesses which would have severely limited the useful life in the field.

The ruggedization pushed the design-related wearout modes way out in time, much more than necessary for the intended environment and also lowered the bottom of the curve as illustrated in Figure 6.2.

Note that this approach leads to a flaw-tolerant design which can be assembled with minor errors and which will still satisfy the requirement of no failures due to the field environments. The

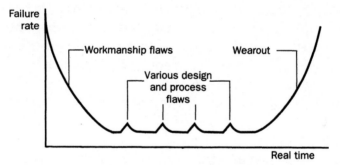

Figure 6.1 Failure rate versus real time for the original product

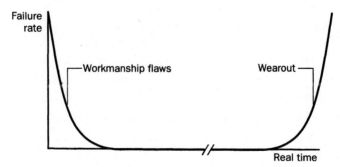

Figure 6.2 Failure rate versus real time for the improved product

124 UNIFORMITY AND REPEATABILITY

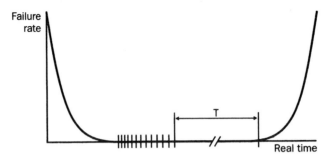

Figure 6.3 Real time to which each position aged the product

concept of the field time, "real time", to which the screen aged the products during screening at each position of the 12-position fixture is shown in Figure 6.3.

During the screen, time compression occurred and so the actual time in the screen is much less than would be required in the field. Real time is shown here for simplification. Note that all positions aged the units under test past the early failure modes, those due to assembly-induced or component flaws. Note also that all the positions do not age the mounted product anywhere near the end of life. Finally, note that if the product had not been ruggedized, then the end of life would probably have been reached at some, or maybe even all, positions on the fixture if a repeated screen had been performed. With the Design Ruggedization approach, now evolved into HALT, it is seen that large variations in the screen damage accumulated during HASS are allowable. This is not the case if small design margins exist, which is normally the case if the advanced design techniques are not used.

Note that if large design margins are not present, design-related failures can be expected since a distribution of the time to failure always exists and some units will fail earlier than the average. This fact is one of the reasons that Design Ruggedization worked so well. HALT works even better today since it is even more advanced and the screening tools, such as precipitation and detection screens, including modulated excitation, are far more advanced than what was in use then. In addition, far superior thermal cycling and vibration equipment is available today than existed then.

A further example taken from switches used on a current

production aircraft is supportive of the above conclusions. A vibration/thermal chamber vendor had developed screens for the switches, performed Safety of HASS, performed screen tuning and then delivered the HALT and HASS system over to the switch vendor. The aircraft manufacturer examined the screens and declared that there was too much difference between the environments at various locations and that the "hot spots" (areas of highest stress) should not be used. The author suggested putting functional switches at each "hot spot" and just leaving them there so that everything was identical to the setup during HASS optimization. After more than 50 HASS profiles had been run, the switches at all "hot spots" were performance checked during HASS and found entirely satisfactory. The "hot spots" were apparently not too "hot".

In a test at Hewlett-Packard [3] in Vancouver, WA, a group of products were run through many HASS profiles on an ink jet printer which utilized a different brand of six-axis vibration system and rapid temperature chamber in an attempt to find out how many profiles it took to reach wearout. After 1,251 HASS profiles, every unit passed a complete Qual test. During production HASS, units screened in various positions had very similar Pareto charts of failures even though the temperatures and vibration levels were substantially different.

All of these case studies point to the conclusion that strict uniformity is not essential *if and only if large margins exist*.

6.3 REPEATABILITY

Each of the 12 screens at Santa Barbara Research Center was quite different in terms of thermal profile and vibration overall level and frequency distribution of the vibration. Since all were equally effective and none took the unit under test too close to the end of life, it then follows that any one screen could be replaced by any other screen and no difference would occur in screen effectiveness. All screens would leave allowable life left in the unit under test. Other programs have shown similar results and it is believed to be generally true that tight uniformity and repeatability are not necessary in HALT and HASS if and only if large margins exist *and are maintained*. The margins can be maintained by using HALT to qualify the margins as time passes.

There are indeed complete families of screens which are effective for any given product *if and only if* the product has significant design margins. This point is extremely important. If a product does not have large design margins, then screening of any type will probably degrade field performance due to premature wearout. HALT is one way of attaining significant design margins at minimal costs.

The benefits of HALT continue to be discovered and this is one of them. If HALT has been properly completed on a product, the product will probably have much more life in it than is required by the field environments. In that case, tight control of HASS parameters is not necessary since the product can afford to lose a substantial portion of the built-in life and still have more than enough for a normal lifetime in the intended use environment. This latitude allows the use of large thermal systems with variations of temperature throughout the chamber and large shakers with substantial variation of vibration across the table surface without concern for the variations. In fact, any six degrees of freedom vibration system will show large variations in the linear accelerations across the table surface due to the angular rotations which cause a gradient in the linear accelerations. Six axis systems exhibit different characteristics, but none is uniform.

6.4 CONCLUSIONS

If HALT and HASS are properly done, tight uniformity and repeatability of vibration and thermal stimulation is not necessary for safe and effective screening. Very substantial variations are allowable if and only if large design margins exist. If large design margins do not exist, a substantial portion of the population can be expected to fail prematurely due to end of life (wearout) occurring. It is essential that Safety of HASS and HASS tuning be performed at every position in the chamber if there are multiple positions as in the examples cited. The same applies to HASS optimization. The successful running of these two tests will show that uniformity and repeatability are not necessary and that the stressing equipment used is safe and effective.

REFERENCES

1. Hobbs, G. K., Triaxial vibration screening – an effective tool, *IES ESSEH*, San Jose, CA, 21–25 September 1981.
2. Hobbs, G. K., Cost effective multiaxial quasi-random stress screening, *Conference of the International Society for Testing and Failure Analysis*, San Jose, CA, October 1982.
3. McLean, H., Highly accelerated stressing of products with very low failure rates, *1992 Proceedings of the IES*.

CHAPTER 7
Physics of Failure

7.1 INTRODUCTION

The study of how engineered products fail is called the physics of failure (PoF). One relationship was very briefly discussed in Chapter 1 and will be discussed further herein. That specific relationship had to do with the mechanical fatigue damage due to vibration, power cycling or thermal cycling; in fact, to any applied cyclic stress that caused internal mechanical stresses. It happens that this particular failure mode is very prevalent in many of the products subjected to HALT and HASS. There are numerous other relationships, so many, in fact, that there is a whole field of study on the physics of failure. For example, the University of Maryland, CALCE Electronic Packaging Research Center, has Masters and Doctoral programs in this field of study with hundreds of students participating. This chapter will be only a brief introduction to the subject and readers interested in a deeper coverage are referred to the references at the end of the chapter.

There are two types of failure. One is from short-term overload and the other is of a fatigue nature. Consider a tensile test specimen. If the load is rapidly increased, an overstress condition will occur where the ultimate strength of the material is exceeded. If a lower intensity load is cycled over some load range, fatigue will eventually occur even for relatively low load levels. Another example is that of current running through a wire. If the current is high enough, the wire will heat up and melt like a fuse. Under lower current, electromigration will eventually remove enough material (usually at a current concentration) that the wire will fail, again like a fuse due to the reduced cross section of wire and the high current density.

In the physics of failure approach, one attempts to determine the underlying mechanism which describes the failure mode with mathematical equations so that the effect of various variables can be studied and the product designed to be more robust and/or the useful lifetime calculated. Note that this does not apply to the concept of MTBF which has to do with a constant failure rate due to flaws rather than wearout, although these too could be studied if one so desired. In order to do so, one would have to assume various flaw sizes and locations in order to attempt an estimate of MTBF. The author views this as rather unproductive labor given the required assumptions. If one had a known flaw and wanted to evaluate its impact on the MTBF, one could do so if the field environments were accurately known, which is usually not the case.

One of the most critical issues in PoF is the need for a rational method to relate the results of HALT and HASS quantitatively to in-service reliability, using a scientific acceleration transform [1]. Elsewhere in this book, the term "time compression factor" has been used instead of acceleration transform. In HALT and HASS programs, the processes expose the weak links in a product and then the PoF approach can be used to analyze the product and look for ways of improving the weak link. If one were to do PoF without the benefit of HALT and HASS, then the analysis would be very lengthy as *everything* would have to be analyzed, not just the weak links because the identities of the weak links are not known *a priori*. This shows one of the benefits of HALT; namely, that it allows a much less cumbersome PoF analysis to be performed. PoF can frequently be omitted if the designer has experience in ruggedizing similar products and knows how to make the product more robust without the benefit of detailed analyses. The author has heard of attempts to perform virtual HALT via the PoF approach. This will not work well for at least three reasons:

1. Design and process defects may not be modeled since they may not be known

2. The analysis would be inordinately long and expensive since all possible failure modes would have to be analyzed in great detail, and
3. The product would be analyzed as it was supposed to be built, not as it will actually be built.

It is much cheaper and faster to build a unit and then test it in HALT to find the weak areas and then analyze only those areas if an improvement is not obvious, which it frequently is. However, before any hardware is built, only analysis (and experience) is available to find weak links and analysis could be used judiciously based on experience to find the weakest links. Experience or becoming "product smart" has tremendous benefits. One of them is knowledge of what not to do and what to do on specific products. The knowledge can be gained by experience, including the results of previous HALTs and PoFs. One does not always have to analyze everything on each new design.

There are three main PoF activities and these are:
1. understand the hardware configuration
2. determine the product life cycle loads in the real-world environments
3. perform an initial PoF assessment of the potential failure modes that could occur.

For example, in order to quantify damage due to random vibration loads, one requires deformation and acceleration time histories at the locations of interest. The spectral density plots will not suffice since one cannot calculate the fatigue damage done by the loading conditions without further information. Note that the time histories are frequently not known, so one cannot perform PoF in the normal manner. One could assume loads and then look for weak links using PoF, however. In this case, only the weakest links would be identified and no lifetime information would be gained.

7.2 MECHANICAL FATIGUE DAMAGE – HOW HALT AND HASS WORK

In HALT, one is attempting to find weak links in the design and manufacturing processes. In HASS, one is attempting to locate the

132 PHYSICS OF FAILURE

flawed assemblies or parts so that corrective action can be implemented. It turns out that the flawed parts or pieces will have a higher stress in them than would a non-flawed similar one and this is precisely why the flawed ones fail. An example would be a solder joint with an air bubble in it or a component lead that had been formed with too sharp a bend or which had been kinked by careless handling. The stress in these areas would be higher than if they had no such defect.

It is a fundamental fact of nature that higher stresses of many types produce failure acceleration factors out of proportion to the increased stress. In most cases, the acceleration is exponential in some form. One such case will be discussed below to clarify some concepts.

One common failure mode is mechanical fatigue due to cyclical stresses caused by thermal cycling, vibration, humidity or even power cycling. Mechanical stress-related fatigue damage follows the relationship [2]:

$$D \approx NS^\beta, \tag{7.1}$$

where

D is the Miner's criterion fatigue damage accumulation

N is the number of cycles of stress

S is the mechanical stress in force per unit area

β is a material property.

Figure 7.1, which is called an S–N (for stress versus number of cycles to fail) diagram is from Steinberg [3]. The graph is derived from tensile fatigue tests on specimens and illustrates that the

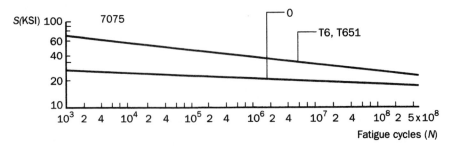

Figure 7.1 S-N diagram for 7075 aluminum

relationship between the number of cycles to fail and tensile stress is exponential in nature and verifies that an equation such as (7.1) exists. The quantity ß is derived from the slope of the curve and ranges between 8 and 12 for most materials in high cycle fatigue; that is, for relatively low stresses and many cycles before failure. S–N diagrams for other materials are similar in nature.

Examining the figure, which is for 7075-T6 aluminum, we get the following points by going across from the applied stress which is the ordinate to the line on the diagram and then going down to the abscissa to find the number of cycles to fail:

1. At 40 KSI (thousands of psi) it takes 2,000,000 cycles to fail

2. At 80 KSI it takes 2,000 cycles to fail.

Therefore, an increase of stress by a factor of 2 causes a decrease in life by a factor of 1,000 times! This acceleration factor is very typical for mechanically induced fatigue and for many other failure modes as well. Note that every cycle of stress does non-reversible fatigue damage which is cumulative and cannot be removed other than by melting the material and fabricating it anew. A few large stress cycles do as much fatigue damage as many smaller stress cycles, and this is why HALT and HASS work as well as they do.

Equation (7.1) also shows why increased stresses accelerate screens so much. Consider increasing the vibration level by doubling the RMS level. If we assume that the beta is 10, then the fatigue damage accumulation rate would be multiplied by some number near 1,000. This means that the screen could be shortened by a factor of 1,000 also. It is this acceleration that leads to very cost-effective screens. In this case, the number of thermal cycles could be reduced by a factor of 1,000 or the duration of the vibration could be reduced by the factor of 1,000. The cost of the equipment to do HALT or HASS is similarly reduced. Time compression by using higher stress levels leads directly to cost compression and that is the foundation of the accelerated techniques.

Parts which fail in service for reasons other than wearout usually have some kind of imperfection which causes an increase in stress. The stress concentrations caused by even a small imperfection may be two or three; i.e. the stress is two or three times as high as in a part without an imperfection. It is seen that even a small imperfection can reduce the fatigue life under some conditions by several orders of magnitude. For example, if the stress concentration factor is 3 and β is 10, then the acceleration factor is nearly 60 million. The stress concentration factor of 3 is rather high but can exist at an abrupt notch or in a poor design with a small radius.

7.3 EXAMPLE OF VIBRATION

Assume that cyclical loading is generated by vibration and that the predominate response which creates the fatigue is at 1,000 Hz. The frequency assumption is not essential to the conclusions and the resulting ratio of cycles to fail is independent of the assumption. The specimen with 40,000 psi applied would fail in 33 min (2,000,000 cycles/1,000 cycles per second/60 s/min = 33.33 min). The specimen with 80,000 psi would fail in 2,000 cycles which would occur in 2 s.

This example shows the value of using the highest possible vibration level which will in turn lead to a higher stress in the assembly. Non-linear effects may invalidate this statement to some degree, perhaps a large one. Doubling the GRMS vibration level would double the product stress (if linearity prevails) and decrease the required vibration screening time by a factor of 1,000 times! This is one of the reasons why all-axis vibration is such a powerful screen and why more flaws are generally found using all-axis vibration than with any other stress.

The cost impact of not using HASS is imposing. For example, if the overall vibration level were to be reduced by a factor of 2, it would require 1,000 times as many shakers to screen the total production. If one shaker costing US$100,000 could screen the total production at a given vibration level, then reduction of the vibration level by a factor of 2 would increase the shaker purchase cost to US$100,000,000! This does not include the cost of the test equipment required at each station, personnel or the new building(s) required to contain the equipment and personnel. The power required to run 1,000 shakers of any type is staggering.

Clearly, one needs to use the most efficient techniques when starting to consider spending extremely large amounts of money. This example could clarify why many companies that try screening using the "normal" methods just give up in frustration. Either the cost is just unacceptable or they try to shorten the screens from an effective time, which may be many hours, down to a few minutes and consequently find very little due to the resulting reduced effectiveness. In this case, failures which might have occurred in five years in the field might occur in a few weeks in the field due to the partial precipitation of the shortened screen, exactly what is not wanted. Either way does not lead to successful competition in today's business climate. It is suggested that the levels be kept up and the time down for rapid, cost-effective screens of any type.

7.4 EXAMPLE OF RATE OF CHANGE OF TEMPERATURE

Another example is taken from [4]. In this paper, Smithson reports on the number of thermal cycles required to precipitate a delamination problem in a transistor. There were 400 000 samples in total and 100 000 were thermally screened in each group run at four different thermal rates ranging from 5 °C/min to 25 °C/min. The test was continued until it was considered that all the weak items had failed, approximately 10%. The number of cycles for the last ones to fail was plotted on linear paper as received by the author. The data were replotted on log–log paper. The results are shown in Figure 7.2 as rate of change of temperature versus number of cycles to fail. The straight line fit of the data implies an equation of the form of (7.1). It is shown that 400 cycles at 5 °C/min would cause the delamination as would 4 cycles at 25 °C/min. An extrapolation to 1 cycle at 40 °C/min was made by the author and is shown in the graph. The results quoted are for a specific defect type and are not generally true for all defect types.

Data from Figure 7.2 can be used to calculate the screen time in hours versus the rate of change of temperature. The two extremes are:

1. 5 °C/min required a 440 h screen time
2. 40 °C/min required 0.1 h.

136 PHYSICS OF FAILURE

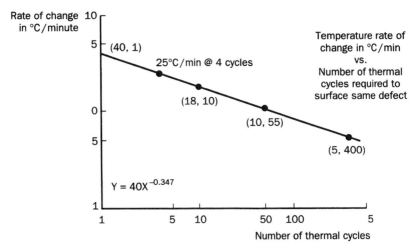

Figure 7.2 Number of cycles required to precipitate the flaws

Notice the factor of 4,400 difference! This means that if one thermal chamber running at 40°C/min can generate a given production throughput, then it would take 4,400 chambers running at 5°C/min to generate the same throughput! If the thermal chambers cost US$50,000, the chamber cost would jump from US$50,000 to US$220,000,000! Again, this estimate does not include the test equipment, people or building(s) required to support the activity, nor does it include the power required. A valid cost comparison must take into account the time or number of cycles required and the power and liquid nitrogen used. Also, note that less total liquid nitrogen is used cycling at 60°C/min than at 25°C/min because of the reduced heat transfer through the walls of the chamber and due to the temperature lag in the chamber skins and other structures during the shorter time period. This effect is even greater if ducting of air to the product exists.

The above example is based on a flaw which is sensitive to the thermal rate, but not all flaws are sensitive to the thermal rate. If the flaws are sensitive to the number of cycles and the total thermal *range*, then the broadest possible range will result in the fewest cycles required and then the screen will be dictated by the time required to run the cycles and the higher rate will give linear, instead of exponential, acceleration with rate. In this case, with a flaw sensitive to the number of cycles and the temperature range, it is seen that total precipitation screen cost is roughly inversely proportional to the ramp rate used. With modern techniques, no dwell time is required since the product temperature, not the chamber air, is controlled during the cycling. Note also that modern screening approaches in use by the leaders in the field combine multiple stimuli in one setup so that the test equipment cost is minimized. One also gains the synergism of combined stresses which precipitate and allow the detection of far more than what an individual stress will accomplish.

One failure mode, that of mechanical fatigue damage, has been discussed so far. This is probably the most frequently occurring mode in electronic equipment and that is why this particular mode was selected for discussion. The field of physics of failure is so large that many excellent books and papers have been written on the subject. No short coverage such as must be done here can treat the subject with appropriate sufficiency. For this reason, a lengthy reference list is given so that interested readers can be directed to sources of further information.

The paper by Ed Hakim [5] covers many failure modes of microelectronics and is certainly suggested reading, as are many of the other references at the end of this chapter.

7.5 WHAT STRESS TO USE

Novices to the field of HALT and HASS often ask the question, "What stress should be used?" The correct answer is, "All of them!" This answer is often not understood without a detailed discussion of the subject since there is a body of literature which professes that burn-in or thermal cycling will do the entire job. In general, those pat answers are completely incorrect since there is no one stress which will expose everything. Also, some flaw types must have combined excitations in order to be found at all. In some cases, it takes a combination of stresses to precipitate a latent defect and another (or several combined) to detect the patent defect.

Several of the more important stresses will be discussed briefly below.

1. *Burn-in* is a condition of elevated temperature, frequently with voltage applied and frequently with reverse bias. The elevated temperature activates chemical reactions and causes some migration effects to take place, both as described by the Arrhenius equation. Some defects discovered by burn-in are: diffusion processes on silicon; oxidation of fractures; and poor timing margins. Very little else is accomplished by this regimen.

2. *Clock variation* may expose poor timing margins.

3. *Power cycling* is turning a product on and off alternately. This causes both a current surge and rapid heating on the junction level, stimulating thermo-mechanical flaws and design deficiencies. Accelerated electromigration is caused by power cycling. Power cycling is particularly effective on power supplies, especially in combination with all-axis vibration and high-rate thermal cycling since the stresses all combine to create very high stresses. A combination to include in a power supply HASS is all-axis vibration, low temperature and power cycling. This combination has been found to be amazingly effective on many flaw types found in power supplies.

4. *Temperature cycling* causes differential thermal expansion and differences in temperature between various parts of a system. The former is not sensitive to ramp rate, but the second one is. Both effects can be present together and are then synergistic. High and low temperatures show up some design marginalities very well. Some defects exposed include: interconnection problems, poor solder joints; defective wire and ball bonds; poor timing margins; and rate-sensitive flaws. Modulated excitation is particularly effective when applied with temperature cycling in order to find cracked conductors. The thermal cycling frequently cracks an item, but the vibration is necessary in order to detect the crack. Modulated excitation is, therefore, an integral part of HALT step stress testing.

5. *Voltage variations* can reveal design margin problems and also marginal performance of specific components. When combined with temperature testing at high and low limits, voltage

variations are quite effective in the detection of marginal conditions, perhaps due to variations of component values as the temperature changes.

6. *Vibration* usually precipitates mechanical design problems, such as large components that are not supported properly, fasteners that are not properly locked in by some means, designs which have ignored the octave rule and cabling problems. Defective solder joints attaching a component to a circuit board are susceptible to vibration, particularly when combined with temperature cycling.

7. *HAST* is a highly accelerated stress test consisting of high pressure, high temperature and high humidity. By utilizing pressures above atmospheric, the humidity can be increased without the formation of water droplets. Humidity may expose poor grounding and isolation problems as well as a plethora of others. This test is regularly used by one auto manufacturer for any under-hood mounted components. Sometimes salt water is sprayed on the product before HAST in order to excite corrosion.

8. *Humidity* is usually run at high temperatures to accelerate corrosion problems. This stimulation is a good one to run after HALT since cracked component bodies may not show up until time and humidity take their toll and cause a long-term problem to show up. It is suggested that all HALT units be exposed to HAST afterward to detect whether any cracked components exist.

9. *Electro static discharge*, or ESD, is used to determine design adequacy and is usually run as a single stimulation since there appears to be little synergism with other stresses. In these tests, high voltages are arced onto the product to determine the susceptibility of the product to ESD. ESD is not usually run on shippable products, but only in HALT.

10. *EMI* susceptibility is often performed to verify design margin adequacy and is usually performed at ambient conditions. Testing done at temperature can reveal problems such as leakage paths not found at ambient conditions.

7.6 VENN DIAGRAMS

The concept of flaw–stimulus relationships can also be shown in Venn diagram form as in Figure 7.3 for a hypothetical but *specific* product. It would be different for a different product. For clarity, not all stimuli are shown. What is shown in Figure 7.3 is the stress which precipitates a flaw from latent to patent [6]. Note that for the hypothetical example given, there are many latent defects that will not be transformed into patent defects by any *one* stimulus. For example, a solder splash which is just barely clinging to a circuit board would probably not be broken loose by high temperature burn-in or voltage cycling, but vibration would probably break the particle loose and there is some chance that thermal cycling would loosen the particle. Also, in order to find that the defect exists, the particle must be seen by eye, heard by sound (as in particle impact noise detection (PIND)) or by the electrical or mechanical function of the device being screened showing some sort of a change. Note that the defect may only be observable during a stimulation and not be observable during a bench test. This is very common and so full power up monitored screening is required to detect patent defects. Many cases have been observed wherein patent defects were only detected under a specific set of screen conditions such as low temperature and modulated all-axis vibration.

Another example would be that of a latent defect of a chemical nature where a high temperature bake would cause a reaction to

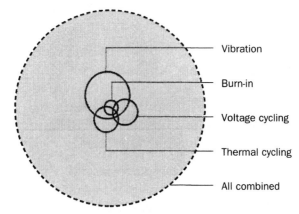

Figure 7.3 Venn diagram for flaw–precipitation stimulus

proceed and a defect to become detectable. Applying vibration, mechanical shock or centrifuging would be completely ineffective in precipitating such a defect. The stimulus *must* be chosen to precipitate flaws into defects that can be found during or after the screen.

Please note that the Venn diagrams have changed over the years and that one drawn in 1981 had a very large circle for burn-in since components were not nearly as reliable then as they are today. The failure rate of standard components today is at least four orders of magnitude better than it was in the early 1980s. Burn-in is still as effective as it was back then, but the population of flaws which can be exposed by burn-in used on a circuit board or higher assembly level is now smaller by four orders of magnitude. The extensive use of burn-in on many products today is misguided and is probably done out of tradition and not because it finds sufficient flaws to justify the costs. Other screens, such as all-axis vibration and high-rate thermal cycling, are vastly more cost effective today than burn-in *on the circuit board or higher assembly level*. For that reason, many leading companies are terminating burn-in in favor of the more effective screens. One must always ask the question, "How can I precipitate the defects and how can I detect them?" Using the wrong set of stresses will not produce results no matter how long (and expensive) the screen is.

The author has found, based on having performed HALT and HASS on more than 100 products of great diversity, that all-axis vibration has uncovered more design and process problems than any other stress used separately. If only one stress could be used, it would be this one. However, it must also be stated that no one stress is "the most effective" since one must always consider the flaw types sought and what stresses are capable of precipitating them and what stresses combined with what tests are capable of detecting them. In general, one must use a number of stresses *simultaneously* for effective screening. Not only is combined stressing vastly more effective than sequential stimulation with various stresses, but combined stressing is far less expensive when the best equipments available today are used.

It is also noted that combined all-axis broad-band vibration and very high-rate thermal cycling precipitate and detect many more flaws when simultaneously applied than when applied separately. There are a *few* cases where separate excitation is better in terms of the fatigue damage accumulation rate, specifically in some surface mount components where the solder fatigue is somewhat inhibited by temperature cycling and vibration alone would be faster. Combined excitation would do the job; it would just take longer. This again says that one needs to understand what is sought and how it fails.

The area in Figure 7.3 that is enclosed with a dashed line but is outside the solid circles is that area that only shows up when combined stimuli are used. The area is very substantial in size since many flaws, when precipitated, may only show up under the combined stimulation. See the section of Chapter 3 on modulated excitations for some examples. HALT results from 47 different products from 33 companies across 19 different industries are described in [7]. The percentages of defects found under each stress are listed in Table 7.1. The order of the listings matches the order in which the stresses were done. The order is very important.

Note that more defects were discovered in vibration than in temperature cycling in spite of the fact that the vibration was done after the temperature cycling. This result is somewhat clouded by the fact that the numbers in the table are where the flaw was detected, not necessarily where it was precipitated. Still, if the HALT were to have stopped after thermal cycling, then the flaws found would only have numbered 35% of the total found. This again shows the necessity of multiple stresses and that thermal cycling is not the most effective screen on a broad-brush basis.

In an unpublished paper presented at a company open house, Dave Jedrzejewski, then with Array Technology, gave statistics

Table 7.1 Percentage of defects found for each type of stress

Stress	%
Cold step	14
Hot step	17
Rapid temperature cycling	4
All-axis vibration	45
Combined temperature and vibration	20

Table 7.2 Stress conditions per numbers of flaws detected

Conditions	%
High-rate temperature transition	12
High temperature extreme	13
Low temperature extreme	29
Combined high-rate temperature cycling and all-axis vibration	46

regarding the type of stress conditions present per percentage of flaws detected in a large (1,400 lb) cabinet of electronics.

An interesting point is that more failures were detected at low temperature than at high temperature. This is a typical circumstance in many electronic and mechanical systems today. Also note that if combined temperature and all-axis vibration had not been performed, 46% of the failures would have been missed.

7.7 CONCLUSIONS

The concept of physics of failure has been described. Relationships between the various stresses and what kinds of flaws are exposed by them has been briefly discussed. A complete PoF approach can provide approximate acceleration transforms [1] as well as helps to evaluate different geometries in a search for a longer life configuration. The PoF approach is currently used by numerous companies. The author thinks that the PoF approach is very beneficial where there is no experience base upon which to draw; however, when such a base does exist and one is "product smart", one may not have to perform PoF. HALT is very powerful in pointing out where the weak links are and thereby very much reducing, or even eliminating, the PoF effort. A sensible combination of HALT and PoF seems to be the best way to proceed. As in most fields, one needs all the tools and skills available in order to be the most efficient.

REFERENCES AND NOTES

1. Upadhyayula, K. and Dasgupta, A., Guidelines for physics-of-failure based accelerated stress testing, *Proceedings of the 1998 Annual*

Reliability and Maintainability Symposium. This paper has many references of interest to PoF. These are all listed below for the convenience of the reader.

2. Miner, M. A., Cumulative damage in fatigue, *Journal of Applied Mechanics*, **12**, (1945).
3. Steinberg, D., Vibration analysis for electronic equipment, John Wiley & Sons, New York (1988).
4. Smithson, S. A., Effectiveness and economics – Yardsticks for ESS decisions, *1990 Proceedings of the IES*.
5. Hakim, E. B., Microelectronic reliability/temperature independence, *Quality and Reliability Engineering International*, **7**, pp. 215–220
6. Hobbs, G. K., Flaw stimulus relationships, *Sound and Vibration*, April 1986.
7. Silverman, M. A., Summary of HALT and HASS results at an accelerated reliability test center, Proceedings of the 1998 Annual Reliability and Maintainability Symposium.
8. Barker, D. B., Vodzak, J., Dasgupta, A., and Pecht, M., Combined vibrational and thermal solder joint fatigue – a generalized strain versus life approach, *Transactions of ASME, Journal of Electronic Packaging*, **112**, pp.129–134, June 1990.
9. Caruso, H., Hidden assumptions in temperature and vibration test time compression models used for durability testing, *Proceedings of the 1994 IES Annual Technical Meeting*, pp. 107–115.
10. Cluff, K. J., Barker, D., Robbins, D., and Edwards, T., Characterizing the Commercial Avionics Thermal Environment for Field Reliability Assessment, *Proceedings of the 1996 IES Annual Technical meeting*, pp. 50–57.
11. Cushing, M. J., Mortin, D. E., Stadterrnan, T. J., and Malhotra, A., Comparison of electronics – reliability assessment approaches, *IEEE Transactions on Reliability*, **42**, December 1993.
12. Dasgupta, A., et al. Material Failure Mechanisms and Damage Models, – A tutorial series containing 14 articles in *IEEE Transactions on Reliability*, lead article appeared in **40**, (1), pp. 531 (1991).
13. Dasgupta, A., Oyan, C., Barker, D., and Pecht, M., Solder creep – fatigue analysis by an energy-partitioning approach", *Transactions of the ASME, Journal of Electronic Packaging*, **144**, pp. 152–160 (1992).
14. Dujari, P., Upadhyayula, K., Dasgupta, A., and Balachandran, B., Applications of wavelets for cost-effective vibration response of electronic circuit card assemblies, Experimental/numerical mechanics in electronic packaging, *1997 Spring Conference of the Society of Experimental Mechanics* .

15. Hu, J. M., Barker, D., Dasgupta, A., and Arora, A., The role of failure mechchanism identification in accelerated testing, *Journal of the IES*, pp. 39–45, July 1993.
16. Jensen, F., Electronic component reliability, John Wiley & Sons, New York (1994).
17. Jensen, F., and Petersen, N. E., Burn-in: Engineering approach to the design and analysis of burn-in procedures, John Wiley & Sons, New York (1982).
18. Lall, P., and Pecht, M., An integrated physics-of-failure approach to reliability assessment advances in electronic packaging, *ASME Journal of Electronic Packaging*, 1993.
19. Larson, T. and Newell, J., Test philosophies for the New Millennium, *Journal of the IES*, pp. 22–27, May/June 1997.
20. Mann, N. M., Schaefer, R.E. and Singpurwalla, N. D., Methods for statistical analysis of reliability and life data, John Wiley & Sons, New York (1974).
21. Meeker, W. Q. and LuValle, M. J., An accelerated life test model based on reliability kinetics, *Technometrics*, **37**. (2), pp. 133–146, May 1995.
22. Nelson, W., *Accelerated testing, statistical models, test plans and data analyses*, John Wiley & Sons, New York (1990).
23. Pecht, M., *Handbook of Electronic Package Design*, Marcel-Dekker, New York (1991).
24. Pecht, M., *Integrated circuit, hybrid, and multichip module package design guideline*, John Wiley and Sons (1994).
25. Pecht, M. and Dasgupta, A., Physics-of-failure: An approach to reliable product development, *Proceedings of the IES*, Chicago, IL, pp. 111–117, August 1995.
26. Pecht, M., Dasgupta, A, and Barker, D., The reliability physics approach to failure prediction modeling, *Quality and Reliability Engineering International*, pp. 273–276, 1990.
27. Pecht, M., Dasgupta, A., Evans, J. and Evans, J., *Quality conformance and qualification of microelectronic packages and interconnects*, John Wiley & Sons, New York, (1994).
28. Pecht, M., Malhotra, A., Wolfowitz, D., Oren, M. and Cushing, M., Transition of MIL-STD-785 from a military to a physics-of-failure based Com–military document, *9th International Conference of the Israel Society for Quality Assurance*, Jerusalem, Israel, 16–19 November, 1992.
29. *Quality and Reliability Engineering International*, **6**, (4), September–October 1990. There are many excellent articles on failure prediction

methodology in this issue, which is must reading for anyone using MIL-HDBK 217-type methods.
30. Rothman, T., Physics-of-failure methodology for accelerated thermal cycling of LCC solder joints, Masters Thesis, Department of Mechanical Engineering, University of Maryland College Park, 1995.
31. Steinberg, D. S., *Vibration analysis for electronic equipment*, John Wiley & Sons, New York (1988).
32. Upadhyayula, K., and Dasgupta, A., Accelerated testing of CCAs under combined temperature-vibration loading, *2nd Annual Workshop on accelerated stress testing*, IEEE CPMT Society, Ottawa, Canada, 16–18 October 1996.
33. Upadhyayula, K. and Dasgupta, A., An incremental damage superposition approach for reliability of electronic interconnects under combined accelerated stresses, *9th Symposium on Mechanics of Surface Mount Assembly*, Dallas, November 1997.

CHAPTER 8
Software HALT: Accelerated Software Coverage and Resolution

8.1 INTRODUCTION

This chapter explains a method to accelerate software development in terms of enhancing coverage and resolution during the software development stage. The methods discussed herein are independent of hardware HALT, the subject of the rest of the book, and are a software/test hardware development tool that should be applied before HALT is begun.

In Chapter 1, the concepts of precipitation, detection, failure analysis, corrective action and verification of corrective action were introduced briefly. Chapters 2 and 3 further discussed HALT, precipitation and detection and some of the details of those techniques. Other Chapters went into other fine points of the methodology. In all of those discussions, the assumption was made that the detection was possible and indeed a fact. This is very frequently not a fact since many faults cannot be detected by the usual test systems which are referred to herein as the "test software", although it is realized that hardware is also involved. Coverage is the property of being able to detect that a flaw exists in a particular location. Resolution is the property of detecting precisely where the flaw is.

In order for either a HALT or a HASS program to work at all, it is necessary to have *all* five items present: precipitation, detection, failure analysis, corrective action and verification of corrective action. That is, without the five, there will be little or no progress on fixing the problems which exist in the product except by experiencing field failures and then going through a long and arduous failure analysis program. If there is no good fault coverage and good resolution, the problems may not be able to be solved even by the best of engineers. Experiencing and rectifying field failures is far too slow and costly a process to allow it to happen in today's competitive marketplace. The accelerated techniques presented in this book address the first two steps in closed loop design and production reliability achievement. However, for detection to be possible at all, coverage must exist.

8.2 DETECTION

Once a latent defect has been precipitated to patent by whatever stress conditions it takes to facilitate this transformation, it is then necessary actually to detect that a patent defect exists. This detection requires that two simultaneous events occur:

1. the defect must be placed into a detectable state, and

2. the detectable defect must actually be detected by built in self test (BIST) or by the test equipment which monitors product behavior during the HALT and HASS processes.

It is noted in passing that attaining a detectable state is somewhat difficult to accomplish, as discussed in Chapter 3 under detection screens. The second event is discussed herein where we assume that the defect is actually in a detectable state while the test system checks the system performance. In this chapter, a detectable state is reached by physically inserting a fault into the hardware, so the assumption of a detectable state is assured.

Note that the probability of detection could be expressed as the probability of putting the system into a detectable state, perhaps by modulated excitation (or by fault insertion), multiplied by the probability of coverage on that particular flaw location:

$$P(D) = P(DS) \times P(COV), \qquad (8.1)$$

where

P(D) is the probability of detection

P(DS) is the probability of a detectable state and

P(COV) is the probability of coverage or simply the coverage since it is a probability.

Note that the first two probabilities are multiplicative and the absence of either one will dictate that no problem will be detected. Obviously, coverage is essential to any HALT and HASS activity.

8.3 COVERAGE

Coverage is defined herein as the possible percentage of defects that the test system in use can actually detect. Many companies report coverage as "about 90%–95%" without any real means of knowing what it really is. Often the coverage is more like 28%–45% based on the results of the application of the automated fault insertion (AFI) techniques utilized at the Proteus Corporation laboratory [1]. This means that more than half of the defects that could occur in a given circuit board or other assembly cannot be detected by the test system!

In this situation of lack of coverage, it is no wonder that many systems pass all of the tests, are shipped to the field and then fail to perform to the customer's expectations almost immediately. Usually, when these non-completely operational assemblies are returned to the factory and the normal tests (with low coverage) used on these types of assemblies performed, no defect is found and they are returned to stock as functioning assemblies. The next customer who receives the faulty (but undetected) item will also experience some performance problem and return the item under warranty. The situation just keeps repeating until a very large "bone pile" of field returns that cannot be verified exists.

Another scenario that frequently happens when low coverage in the test system exists is that an assembly will go into service and then fail for some unknown reason. The reason might be that a solder joint broke during shipment or was broken during the installation, maybe by rough handling. When the assembly is returned to the factory and tested, no defect will be found. This is not surprising since the test system may not be capable of detecting the flaw even if it is in a detectable state and the unit is truly defective. This unit would be reused with the same result again. These two scenarios have been seen many times by the author and that is why this chapter is included in this book. Both of these scenarios are readily cured with good coverage and resolution and with good detection screens.

In one documented [1], but company private investigation, it was found that the coverage was only 7% although the software engineers had estimated a coverage of 95%! Such low coverage is a catastrophe just waiting to happen for both the manufacturer and the end user.

HALT is frequently run on a new product using a very early test software version. Some detectable defects will, of course, be missed. Most mechanical failures will be found as will some electrical ones and so performing HALT on early development hardware is beneficial because of the time lead gained by the discovery of any defects. As the hardware and software develop, more HALTs should be performed. It is frequently very beneficial to retain early HALT hardware units and then to perform modulated excitation on them whenever new software with better coverage and resolution becomes available. This approach may expose some faults not discovered before due to lack of coverage.

8.4 RESOLUTION

Resolution is the ability of the test system to determine where the defect actually is once a defect is known to exist by lack of performance of some type. For example, the test software may give an error code which does not allow accurate location of the defect, but only gives a symptom such as "no output from device a". This would be symptomatic of poor resolution. If, however, the test system were to give an error code that was very explicit as to location such as "No output on pin x of device y", then that would be much better and one would then know to examine device y

further, particularly those features which should result in an output on pin x of that device. A very crude test of extremely poor resolution would be a "go–no go" test. It passes or it does not, and no information as to why it does not pass is given. In complex systems, this would be of little use for fixing the product.

8.5 SOLVING THE DILEMMA [1, 2]

The problem of a test system that cannot detect detectable defects can be addressed in digital electronics by utilizing at-speed functional testing combined with fault insertion. At-speed is necessary because, as technology advances, the operating speeds of state-of-the-art designs continue to increase. With these increasing speeds, a larger percentage of the defective products escaping from the manufacturing line will be defective due to excessive signal delays. Thus, it is important to apply a test which will detect as many delay defects as possible. In some cases, design verification vectors can be applied at-speed in order to determine whether the product operates correctly at the rated operational speed. However, it has been shown that design verification vectors are not necessarily effective at detecting manufacturing defects since they were not generated to target them [3, 4]. It is therefore seen that it is important to use a test that was specifically developed to be applied at-speed in order to test for the delay effects that grow in significance for high-speed circuits. The fault coverage of an at-speed functional test is frequently unknown and it is important to know what the fault coverage is in order to know how effectively it will detect faulty products. For tests which are designed to find delay defects, it is desirable to understand the delay defect coverage. However, for an at-speed functional test, this is a difficult task. The stuck-at fault coverage of an at-speed test has been found to be a good indicator of performance for an at-speed functional test. It has been shown that the stuck-at fault model may be inaccurate for modeling real defects, but a sufficiently high coverage single stuck-at fault model may be adequate to achieve a high level of quality [3, 4]. Therefore, a sufficiently high stuck-at fault coverage will imply that the test coverage for other types of faults is high as well.

Fault insertion is accomplished in several variations and each will be discussed briefly.

8.6 SOFTWARE FAULT INSERTION

Software can be used to inject faults. This method has several problems, some of which are:

1. the analysis of a system assumes that the system is built as envisioned by the designer which may not be true if there is a fabrication fault
2. the software itself may have bugs in it and
3. the software cannot access all locations that may have a fault and may not be able to provide stuck-at or bridging faults.

8.7 SIMULATED FIELD EXPERIENCE

Multiple systems can be run in simulated environments gaining "field experience" of *real* faults if there are any in the units selected for the extensive tests. There is no time compression at all in this approach and so it is an extremely expensive method to find faults. In addition, some faults may be intermittent in nature and cannot be captured for analysis. Something that cannot be captured cannot be examined, so no progress is made in solving a problem.

8.8 FIELD EXPERIENCE WITH CUSTOMER USE

Field experience can be used to find faults. No time compression at all is extant. This method in general is probably the worst way to find faults in a new system. System failures may cause severe problems for the end user as well as for the manufacturer.

8.9 MANUAL FAULT INJECTION

The physical fault injection process is simple, but very tedious and time consuming. Faults that can be manually injected in digital hardware include net stuck-at faults, pin open faults and pin bridging faults. In this procedure, the first step is that the product under test is powered down. The fault is injected, the power is turned on and then the test suite is run to determine whether the fault can be detected. If the functional test executes and fails, then the fault has been detected and any diagnostic information is logged by the person running the tests before powering down the unit under test. Next, the fault is removed and the tests re-run to determine whether normal operation has returned. If so, the fault injection probe is moved to a new location and a new fault is injected there. Then the unit is powered up and the test suites run again to determine whether the fault can be detected. This procedure is run over and over again until all practical sites for faults have been examined. There may be thousands of sites on one circuit board. Obviously, this procedure takes a long time and is susceptible to errors if done by a person, no matter how dedicated. In addition to the time required, manually injecting faults on a fine pitch card can be very error prone and potentially damaging to the "golden" card.

When a fault that has been inserted is not detected, this fact gives the software engineer information that may allow a software improvement so that the fault can be detected. In the process, both coverage and resolution can be improved.

Even if a human who would never make a mistake could be found, manual fault injection is not viable in today's rapid product development scenarios due to the long time required to do the work. A typical board might require 90 days to inject faults manually.

8.10 AUTOMATED FAULT INJECTION

AFI uses a robotically controlled probe to inject faults into the unit under test (UUT), be it a card or a system. The UUT's computer aided design (CAD) database is used to obtain the coordinates of the appropriate places to inject faults. Upon placement of the probe at the correct spot for fault injection, the probe tip is engaged, making contact with the UUT. Then, an appropriate fault is injected and the robot signals the functional tester that the UUT test should be run. The tester runs the tests and notes whether any faults are found and what the diagnostic codes are. Upon completion of the test, the functional tester tells the robot to remove the fault and the tester performs the test again to ensure recovery of the system. Then the tester tells the robot to move to another spot and again inject a fault. The robot then moves to another location, engages the probe and the process is repeated until all appropriate locations have been injected with appropriate faults and the tester has run the tests and noted all diagnostic information.

The sequence is depicted in Figure 8.1 [5].

The AFI used for the generation of information used in [2] was the DVT-100 by Proteus Corporation [1] shown in Figure 8.2. Since it is the only currently available robot known to this author for such injection of faults, its capabilities will be mentioned here. The DVT-100 has an automated fault injector which allows the programmer to select the type of fault to be injected and where it is to be injected. Stuck-at zero and stuck-at one faults are typically injected for fault coverage determination and for a fault dictionary (list of faults generated and what the diagnostics said) generation. Stuck-at means that the value at a particular point is held to some constant value. Voltage levels and current limits can be specified. The system also provides for an externally generated signal to be injected at a particular point on the UUT. Fault timing is important and the DVT-100 has several triggering mechanisms. Injected faults can be triggered by program control or by the rising edge of an externally supplied signal. The fault can be injected immediately or delayed if programmed with a delay. The time during which the fault is injected is also controllable. Contact with the point at which injection is to be accomplished is assured by a mechanism which positively assures contact, thus avoiding a false escape. The accuracy of placement is ± 0.0005 inches.

In general, many different types of at-speed functional test setups can be used with AFI. The types that are currently in use range from PCs, cards cabled to controllers which supply the test stimulus, cards with self-test cabled to power supplies, disk drives and many other types of setups. In all cases, the functional test controller communicates with the AFI tool to sequence through the nets of the card for fault injection.

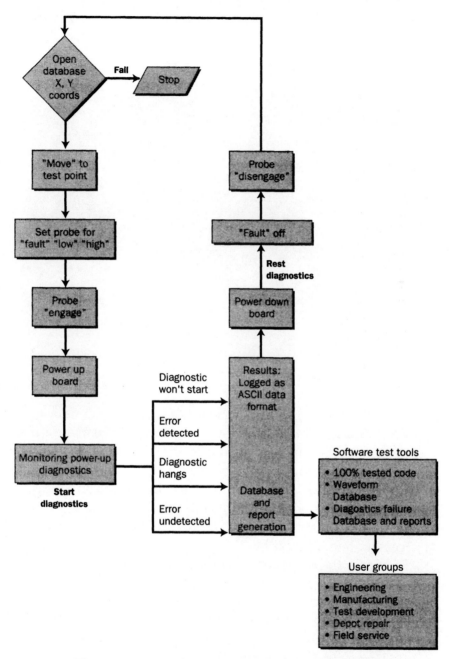

Figure 8.1 A typical automatic fault injection sequence

156 SOFTWARE HALT

Figure 8.2 A robot fault injection system (Courtesy of Proteus Corporation)

The raw data obtained from the tests consist of net names, fault values, pass/fail information and captured diagnostic information. Post processing can provide the desired fault coverage information. Even further post processing can provide a dictionary of fault/diagnostic messages for use in troubleshooting production hardware which exhibits a fault. The fault dictionary can also be used to analyze the diagnostic resolution and to improve it if necessary and if possible.

AFI allows the designer to evaluate fault tolerant mechanisms such as reconfiguration and recovery schemes to determine whether they will work. One can also evaluate performance loss in the event of a fault. All of AFI is based on using real hardware with real potential faults. The same results occur in the AFI tests that would happen in the real world.

The AFI technique discussed here is valuable for obtaining fault coverage information and diagnostic data in an accurate, rapid and cost-effective manner. This method is now required of its vendors by many manufacturers. The rapid product development cycles extant today demand it. In one actual case, a test done manually took four weeks while when done with AFI it took only one weekend. This, then, relates to faster software/test system development and leads directly to faster and better HALT which, in turn, leads to faster and better hardware development. The method is a forced or accelerated software development tool and this is why the name of Software HALT was used.

8.11 SHIPPED DEFECT LEVEL IN PRODUCTION

Measuring the fault coverage and then improving it will decrease the shipped defect level in production significantly. It will also substantially reduce the scrap and rework costs as the inevitable problems which occur in production can be more readily solved in terms of both time and cost. Field returns with no defect found will also be drastically reduced as a result of fewer shipped defectives and also the ability to detect those that really have failed.

We can get an estimate of the change in shipped defect level by using the well-known model of Williams and Brown [6] who published in 1981 a model which demonstrates the relationship between manufacturing yield, Y, single stuck-at fault coverage, T, and defect level, DL. If one assumes that only stuck-at faults occur and that the probability of any particular stuck-at fault occurring is equiprobable and independent, then the defect level can be approximated by

$$DL = 1 - Y^{(1-T)} \qquad (8.2)$$

Although this model makes many simplifying assumptions, the historical data have shown that it effectively demonstrates the trends and relationships between defect level, fault coverage and yield.

We begin by using the model to estimate the effect of improving the net level single stuck-at fault coverage of a card or system. Let Y be the first pass manufacturing yield and let T_i be the measured stuck-at fault coverage. Then the model shows that:

$$T_i = (N_0 + N_1)/N_T,$$

where $N_{0(1)}$ = is the number of stuck-at-0(1) faults detected during fault injection and NT is the total number of stuck-at faults injected. Note that N_T could be equal to $2N$, where N is the number of nets on the card or system, excluding power and ground. However, on some nets, only one type of stuck-at fault makes sense. A net with no signal, however, that is pulled down to ground through a pull-down resistor, would only need a stuck-at 1 fault injected. In this case, a stuck-at 0 fault is undetectable and is deleted from N_T.

Let T_1 be the measured stuck-at-fault coverage of a test set. Then,

$$DL_1 = 1 - Y^{(1-T_1)}$$

Now, suppose that the stuck-at fault coverage of the test set is improved to T_2, measured by whatever means are possible. Then,

$$DL_2 = 1 - Y^{(1-T_2)}$$

The defect level improvement from T_1 to T_2 is:

$$\Delta DL = DL_1 \_ DL_2 \qquad (8.3)$$

By multiplying the defect level improvement by the number of units tested, one has a prediction of the number of defectives that were prevented from shipment to the customer. This is the result of changing the fault coverage of the tests.

An example may clarify the magnitude of the situation. Let us assume several yields and several coverages just for an example. We can calculate from Equation (8.2) the shipped defect levels as in Table 8.1:

Table 8.1 Shipped defect levels

Yield Coverage %	50%	95%	99%
50	0.2900	0.02500	0.00500
95	0.0340	0.00260	0.00050
99	0.0069	0.00051	0.00010

From this example, we can see that if we have excellent yield; that is, very few mistakes in assembly, coverage makes only a small, but perhaps very significant, difference in shipped defect levels. However, if the yields are somewhat lower, nearer to reality, high coverage makes a very large difference in shipped defect levels. Obviously, the better the coverage and the better the yield the better the field reliability will be. However, overriding all of this is the fact that if the coverage is poor, then not many design defects will be found in HALT since every unit tested can have the same problem and, if the test setup cannot detect the problem, none of the units will show a fault and no design fix will be implemented! Sample size will not save us here since, in this case, the test software/hardware just cannot detect the failure to perform. This is one of the real benefits of the better coverage during HALT to find design defects and process problems which are location dependent. In HASS, the above analysis of shipped defect levels would apply.

Yield shows the benefit of corrective action in production; that is, good corrective action will lead directly to higher yield and that leads directly to fewer field failures regardless of the coverage. This comment does not apply to HALT where no corrective action will occur unless a problem is detected and that requires coverage at least where the specific defect under consideration is located.

The manual version of fault injection may take several months to perform and is fraught with potential for mistakes due to its tedium and duration. The unattended automated variety may take a few days or a just a single weekend.

AFI can be used during development to obtain better fault coverage and resolution. This will lead to a much more successful HALT since most, or hopefully all, design defects and marginalities will be discovered and the design changed to eliminate them early in the design phase. The automated feature allows operation of the tests 24 hours per day without stopping so that the data are obtained rapidly and cost effectively. A dictionary of fault versus failure diagnostics can be made up that will aid in troubleshooting design, production and field failures. When changes are made to the product or to the software/test equipment, the automated test approach can be utilized quickly to retest the diagnostic performance for improvement or to ensure that there is no degradation in coverage and resolution. Test consolidation can be accomplished by using the data gained in the automated tests. The time necessary to develop the diagnostics can be reduced by accurately knowing how good the diagnostics are at any time. Quick correction of design problems requires quick discovery that a problem exists. Adequate test coverage is therefore at the heart of many corrective actions and is therefore crucial to HALT and HASS on electronic systems.

In production, the improved test coverage makes the product shipped much more reliable in the field since very few patent defects will escape the tests run because the improved coverage will ensure few escapes. Production ramp to full scale will be faster since the existing problems in production processes will be found much more quickly with the improved coverage.

Field results are also vastly improved when a high coverage bank of tests is in place since most, or hopefully all, production problems will be discovered and fixed using the combination of HASS and fully monitored tests with high coverage. For the few units that do fail in the field, the frequency of no-fault-found will be greatly reduced. It has been typical in the past to have about 50% of the field returns not fail when tested in the factory or at the rework facility. Excellent coverage combined with a proper detection screen will make a significant improvement in this ratio. When used in conjunction with good detection screens, the no-fault-found number should drop to near zero, unless the product never really did fail in the field use.

Automated tests using fault insertion give the best possible assurance that products have been tested as thoroughly as is humanly (robotically) possible. Anything that improves the detection, failure analysis and corrective action will significantly affect production costs, field reliability and, ultimately, profits.

Obtaining the highest possible test coverage is necessary in HALT and one of its benefits in HASS is that it can allow one to have the knowledge to reduce comfortably the screens to a sample. AFI at this time is an emerging technology used by only a few of the leaders in reliability techniques. Further expansion of the technology in its current application, digital electronics, is expected as we gain knowledge and experience with the methods. AFI could possibly be used on other types of hardware as well.

8.12 AUTOMATED SIGNAL INTEGRITY TESTING

In addition to AFI discussed above, the robot used for that task can also perform automated signal integrity studies (see Figure 8.3). Signal integrity is somewhat loosely defined as "proper signal rise/fall times, absence of ringing or excessive overshoot or undershoot, absence of 'glitches' and 'reflections', etc". Analyses of logic "signal integrity" have for a long time now been in the engineer's tool kit for the evaluation of prototype designs. HASS processes are designed to exacerbate latent design, component and/or process defects by creating conditions which cause actual failures in the unit under test. Signal integrity evaluation offers the possibility of detecting latent defects which may not be exposed as actual malfunctions during relatively brief HASS due to the highly marginal and intermittent nature of problems caused by poor signal waveform quality. Thus, signal integrity analyses can be viewed as an extension of HASS methodologies.

CAD simulation tools exist which can help detect some PCB layout problems which might cause signal integrity problems, but these tools do not guarantee performance in actual operation. All analysis approaches analyze things as they are supposed to be, not

Figure 8.3 A dual probe signal integrity tester next to a US ten cent coin (Courtesy of Proteus Corporation)

as they actually are. This is a basic failing of analytical approaches. Manual analysis of a design is a very time-consuming task requiring the skills of a very experienced engineer. When all the combinations of components from various vendors and other combinations are considered, the task takes on enormous proportions. Therefore, these tools are frequently not used.

However, with the robotic system available for AFI, one can have the robot look for signals of questionable quality and then have the design engineer examine only those selected for further scrutiny by the robot/test equipment team working together. This is good use of equipment time as well as that of the design engineer. This application is just in its infancy and more developments are expected to occur in the future.

8.13 TIME DOMAIN REFLECTOMETRY (TDR)

As computers become faster and faster, the constraints on memory board design become much more stringent. One computer manufacturer is requiring that 100% of all traces on memory boards be checked for resistance. The variation allowed is currently only 10%. This will probably tighten up as the computer speeds increase. The only conceivable way to measure 100% of the board traces is robotically. The robot mentioned above has that capability. This application is mentioned for completeness as it has become essential on memory boards for the faster computers of today. TDR does fit in with HASS because it is a tool for determining when corrective action may be necessary.

8.14 CONCLUSIONS ON SOFTWARE HALT AND SIGNAL INTEGRITY TESTING

The techniques of AFI and signal integrity testing using robotics are revolutionizing electronic design and software development. Several of the world's leaders in computers are using the techniques, but are not publishing openly due to the very large advantage that the methods supply. Good coverage and resolution are essential to a successful HALT and HASS program.

The activities described in this chapter are not a part of hardware HALT, but are done before hardware HALT and are software and electronic test development tools. Since software is frequently late in many programs, the effectiveness of HALT may be somewhat to severely reduced by the late addition of good coverage. Early HALT units can be retested when new software is available by simply performing modulated excitation and looking for failure modes not detected in the previous tests even though it may have been patent.

NOTES AND REFERENCES

1. Proteus Corporation, 2875 S. Tejon, Englewood, CO 80110, 303-762-6426, www.ProteusDVT.com
2. Stewart, B., Fault coverage and diagnostic efficiency related to accelerated life testing, *Proceedings of the 1996 Accelerated Reliability Technology Symposium*, Hobbs Engineering Corporation, Denver, CO, 16–20 September 1996. This chapter has drawn heavily from this reference with permission from Dr Stewart.
3. Franco, Farwell, W. D., Stokes R. L. and McCluskey E. J., An experimental chip to evaluate test techniques: Chip and experiment design. In *Proceedings of the International Test Conference*, Washington, DC, pp. 653–662, 21–26 October 1995
4. Ma, S. C., Fratico, P. and McCluskey, E. J. An experimental chip to evaluate test techniques: Experiment results. In *Proceedings of the International Test Conference*, Washington, DC, pp. 663–672, 21–26 October 1995.
5. Lucas, B., Automated fault injection speeds resolution of design quality issues, *IEEE Robotics and Automation Magazine*, March 1996.
6. Collins, P., New tools for test cost reduction, *European Design and Test Conditions User Forum*, Paris, France, pp. 251–255, 11–14 March 1996.

CHAPTER 9
Equipment used for HALT and HASS

9.1 INTRODUCTION

HALT and HASS are used on a variety of products. The equipment used to apply the accelerated stresses is therefore very diverse and is sometimes very specialized. The emphasis will first be placed on electronic hardware. This is done for several reasons, among them the fact that HALT and HASS have been applied more to electronic products than to other products and that a generic discussion on electronic products can be done. For other products, one is immediately engulfed in product-specific functional issues that may cloud the conceptual matter.

The selection of vibration and thermal equipment to perform HALT and HASS will affect the outcome in terms of expense and effectiveness to an enormous extent. The mistake made most frequently by the novice is to try to use single-axis electrodynamic, mechanical or hydraulic shakers and mechanical refrigeration thermal systems, collectively called the "classical equipment" by the author. The usual selection of this ineffective equipment is based on the ready availability of the "classical equipment" that has been around for decades and is present in prolific numbers at bargain prices on the used market. The reason for the bargain prices is that most military contracts no longer call out the MIL-SPEC types of tests and so much of the equipment has been surplused.

It is just natural to turn to the "classical equipment" when starting to do screening. The author did just that himself when he first started screening in 1979. However, a fortuitous requirement by management (based on company politics) to use a new all-axis shaker for production screening led to the immediate discovery of three design flaws which had not been detected in the Design Ruggedization using sequential single-axis excitation on a single-axis electrodynamic shaker [1]. This one experience changed the author's opinion from apprehension to appreciation of the then new and strange shakers. Since then, the all-axis shakers have been the systems of choice and the author has designed four different types of them. One fundamental limitation of the early impact shakers was a very low level of excitation in the critical low-frequency range, a feature which is now available in the more recently developed equipment which had input essentially down to 1 Hz. It was found that the all-axis shaker seemed to do the screening job much faster than the classical equipment, but the technical reasons for it were not fully appreciated at that time. Today, the probability density, the spectral density and the all-axis motion of the shaker are known to be the source of the improved efficiency of the all-axis shakers. These will be discussed later.

9.2 VIBRATION EQUIPMENT

The purpose of a vibration screen is to excite all the modes of vibration of a system to a level high enough to precipitate flaws in design or fabrication. The real world; for example, a truck ride, can be six-dimensional containing three linear accelerations and three angular accelerations. The angular accelerations, using aircraft terminology, are pitch (nose up and down), roll (wings banked left and right) and yaw (nose pointed left and right). These angular accelerations are able to excite modes that single-axis shakers cannot excite at all. In addition to the above considerations of exciting all modes, it is also imperative to excite all modes *simultaneously*. The last statement has been proven hundreds of times over in the last few decades as the argument raged as to whether to use swept sine or random vibration. Random, of course, won out based on exciting all modes excitable *in a single-axis* simultaneously. Here, too, the probability density is critical to the comparison. The newest all-axis shakers excite all six degrees of freedom over a very broad frequency range and excite all modes that can be excited by field environments, a significant advance over the now obsolete linear single- and dual-axis shakers. One new shaker has the capability to control the spectrum in the lower frequency range. See Table 9.1.

Note that if all of the modes that the real world will excite are not excited in a screen, then the real world may, and probably will, precipitate defects that the design and process screens missed. This fact makes it very important to use six-axis shakers in design and process screens since it is desired to find all of the flaws that will show up in the field environment.

9.3 VARIOUS VIBRATION SYSTEMS

First, some definitions need to be made. Sinusoidal vibration is measured by the amplitude of the signal in gravity units or Gs. Random vibration is measured in terms of the power spectral density (PSD) which is in the frequency domain in dimensions of $(GRMS)^2/Hz$. In order to obtain the PSD, a narrow band analysis of the signal is done as represented in Figure 9.1.

In order to quantify fatigue accumulation, a time history of acceleration is necessary. However, this is frequently not available and so the probability density is used to make a probabilistic estimate of the fatigue damage. The probability density represents the distribution on acceleration from the mean. It is physically measured by time sampling at a fixed frequency at least three times above the desired acceleration bandwidth and then the number of occurrences at each amplitude is plotted giving the probability density of the acceleration. The probability densities of all the shaker systems listed in Table 9.1 are not available as of this publication. Also, some of the systems have variable spectral densities by design, so no accurate plots can be published. For that reason, only a general discussion of the probability densities will be given.

From Equation 1.1 (where S was a stress) it can be seen that a few high acceleration cycles (which result in correspondingly high

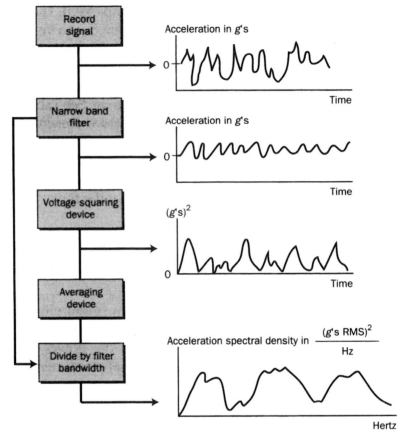

Figure 9.1 Power spectral density determination

stress cycles) generate substantial fatigue damage. Therefore, shakers that produce very high deviations from the mean will generate fatigue damage much faster than shakers that exhibit lower deviations from the mean. Electrodynamic shakers have traditionally been restricted to 3σ in military testing programs in the United States. This tradition severely restricts fatigue generation. The damage prevented by cutting off the signal at 3σ is more than what is left after cutting off the signal! It is no wonder that many military items passed the test but failed in service since the real world does not provide a convenient cutoff at 3σ. Various impact-type shakers have different probability densities and should be investigated individually.

Table 9.1 Comparison of Vibration Systems

Type of shaker	Axes	Spectrum	Frequency (Hz)	Stroke (cm)
Mechanical	1	Sine	Low	2
Hydraulic	1–6	Any	Low–Med	Varies, can be large
Piezoelectric	1	Any	500–20,000	0.1
Electrodynamic	1–6	Any	10–5,000	2–5
CUBE™ system[1]	6	Variable	0–250	5
Accelerated Stress Test[2]	6	AST	0–10,000	2
OmniAxial[3]	6	OmniAxial	0–10,000	2
Modular™ vibration system[4]	6	Variable	0–10,000	3
Premier™ Vibration System[4]	6	Variable	0-10,000	Varies, can be large

Note:
[1] The CUBE™ 6 Degree of Control™ Vibration Testing System is trademarked by Team Corporation.
[2] Thermotron uses the name AST for their system.
[3] The name OmniAxial is used by QualMark Corporation.
[4] Modular™ Vibration System and Premier™ Vibration System are trademarks of Hobbs Engineering Corporation.

Electrodynamic, hydraulic and piezoelectric shakers can all be grouped together in terms of the probability density of the vibration since all use similar control systems. The shakers are not at all the same in terms of other units, such as force rating, frequency range, maximum displacement. See Table 9.1 for some general features of the various shakers.

The mechanical vibration systems excite one direction at one frequency. In HALT and HASS applications, this is nearly worthless and so the mechanical systems are never used to the author's knowledge. The simplest and least effective shaker is the single frequency or swept frequency mechanical shaker. There are various forms of mechanical shakers including inertially excited ones using rotating masses to create sinusoidal forces in a given direction and ones that move the shaker table utilizing a rotating shaft and a connecting rod. Both create a probability density that never exceeds the square root of twice the GRMS level.

The hydraulic vibration systems have vast power; i.e. force ratings, and so are suitable for heavy packages at high acceleration levels. These shakers and the pneumatic shakers as well have limited frequency response due to the compressibility of the fluid used and this is what limits the frequency response. The single-axis hydraulic vibration system has been used by the author to excite large racks of electronics vertically to investigate what might fail in a violent truck ride where most of the accelerations are vertical.

The piezoelectric shakers are very small in displacement and so have little low-frequency acceleration. They have been used by the author to perform a six-axis vibration on an M1 tank thermal gun sight up to 20,000 Hz when investigating pressure bonds on jumper leads which had resonances in the 12–14 kHz range. The production vibration HASS level was 40 GRMS in the early 1980s.

Electrodynamic vibration systems have several benefits, including excellent spectrum control, but this is not necessary in HALT and HASS. Drawbacks include cost, reliability and the substantial magnetic field near the shakers. Many electronic products are microphonic in such an environment and so cannot be monitored during excitation, making use of the electrodynamic systems for HALT and HASS problematical at best. The six-axis varieties of the electrodynamic shakers are very expensive, ranging in the multiple millions of dollars. The reliability of the six-axis electrodynamic systems is far below that of the competing six-axis systems.

The CUBE™ is a true six-axis low-frequency system used in automobile and other applications. It has spectrum control and it can perform time history replication if desired as can all of the hydraulic systems. Items can be mounted on the sides as well as on the top of the system. Figure 9.2 illustrates the excitation of an automobile door and Figure 9.3 illustrates the excitation of a complete seat assembly.

The all-axis shakers which utilize an impact excitation such as the AST, the OmniAxial and the Modular systems generate the highest peak accelerations and therefore have the longest tails in the probability distribution at the higher and lower standard

Figure 9.2 The CUBE™ system exciting an automobile door (Courtesy of Team Corporation)

deviation levels. It has been found [2] that all types of impact shakers generate fatigue damage (in their respective frequency ranges of excitation) in the unit under test about 10,000,000 times as fast as the true random shakers (electrodynamic) when both types are run at the same RMS G level. It is suggested that one use the terms "sinusoidal Gs" for the mechanical shakers or for the electrodynamic or other shakers when running sine, "random G's RMS" to describe random vibrations of the electrodynamic, piezoelectric and hydraulic shakers and note the PSD or the brand and model number on the all-axis shakers because their behaviors are distinctly different. All of these shakers are truly very different in all measures of behavior, and to class them all together in terms of Gs RMS is technically extremely incorrect and very misleading. One should specify the number of axes, the spectra, the probability density and the GRMS in order really to specify what vibration was used. A large high rate thermal chamber combined with an all axis vibration system is shown in Figure 9.4.

172 EQUIPMENT USED FOR HALT AND HASS

Figure 9.3 The CUBE™ system exciting an automobile seat assembly (Courtesy of Team Corporation)

Another system, this one of moderate size, is shown in Figure 9.5 as built by QualMark Corporation.

The Modular vibration system is one of modularity as the name implies. By varying the fixturing and interface material between the fixture and the shaker table, the vibration spectra can be varied over quite broad ranges from low frequency only to very broadband, 0–10,000 Hz. In addition, some spectrum shaping can be done.

VARIOUS VIBRATION SYSTEMS   173

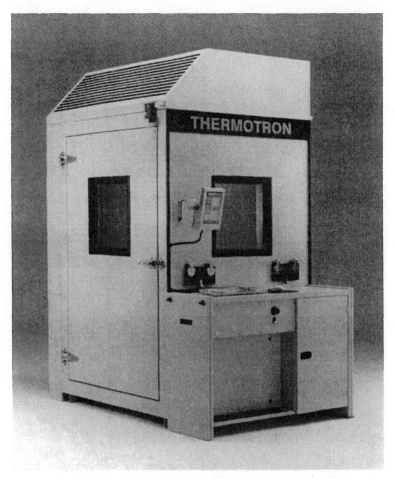

Figure 9.4   A large combined stress chamber, an AST (Courtesy of Thermotron, *A Venturedyne, Ltd. Company*)

The Premier system is a combination of a hydraulic or a pneumatic shaker to obtain the low frequencies and an impact shaker to obtain the high frequencies. This system is fully controllable in the low-frequency range. The Modular and the Premier can be used together to obtain nearly complete control of the entire spectrum if so desired.

An aircraft structural test set up is shown in Fig. 9.6. Hydraulics are normally used for the very low frequency loading. These tests are a close relative of HALT.

174 EQUIPMENT USED FOR HALT AND HASS

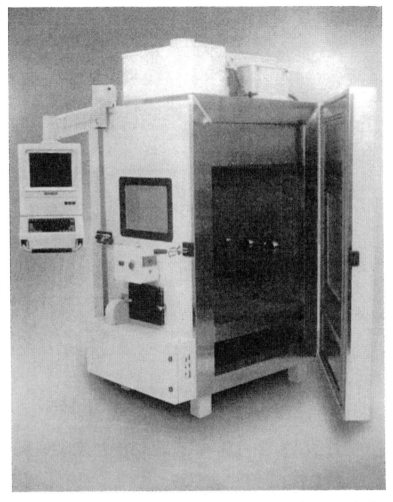

Figure 9.5 A moderate-size combined stress system, the OmniAxial vibration system (Courtesy of QualMark Corporation)

9.4 THERMAL EQUIPMENT

Thermal cycling or burn-in equipment comes in two basic varieties: compressor (mechanical refrigeration) cooled and liquid nitrogen cooled. The mechanical refrigeration is similar to home air conditioning and is very efficient when steady state temperature or very slow thermal cycling is desired. However, when high-rate thermal cycling is desired, the liquid system is more efficient

THERMAL EQUIPMENT   175

Figure 9.6 An aircraft structural test (Courtesy of Cessna Aircraft Company)

in terms of cost to run and is capable of rates of change far above those of the compressor systems. Table 9.2 shows some properties of interest for both types of systems.

Purchase cost is self explanatory.

The installation cost must be carefully examined since several subtle items crop up. The liquid systems are most efficient when equipped with vacuum jacketed lines that cost in the neighborhood of US$125 per lineal foot (in 1999). The mechanical refrigeration systems require cooling water and lots of it. If systems of the mechanical variety are used, a new cooling tower may be required. It generally works out that the two systems cost just about the same to install when everything is taken into account.

Operating costs are as shown *if and only if* a best comparison is done; that is, if each system is used in the optimum way considering its capabilities. Many novices make the mistake of comparing a mechanical refrigeration system, running its most efficient cycle, to

Table 9.2 Comparisons of Thermal Systems

Feature	LN$_2$	Mechanical refrigeration
Purchase cost	1×	2.5×
Installation cost	About the same when installation is included	
Operating cost	1×	6×
Maintenance cost	1×	400×
Reliability	Excellent	Relatively poor
Size	1×	2×
Weight	1×	2×
Noise	55 dBA	70–85 dBA
Thermal range	−100 °C to 200 °C	−40 °C to 177 °C Cascaded to −65 °C
Product ramp rate	60 °C/min	30 °C/min

a liquid system running the same cycle as the mechanical system. This is a totally invalid comparison because the liquid system can run over a much broader temperature range at vastly higher rates and, therefore, only one or a few cycles are needed for an effective screen. Each system's best performance cycle should be considered in order to get a valid comparison of operational costs. Although many companies neglect the cost of power, it is certainly significant and must be counted in the total costs. The author has seen many companies totally neglect the power costs and come to a financially extremely unfavorable conclusion as the power consumed may cost hundreds of thousands of dollars per year per chamber. One company reported to the author that running their compressor-based system for one year cost them US$179,000. When the company switched to liquid nitrogen, the total energy costs dropped to US$49,000. When a company may have many chambers running HALT and HASS, the difference can be very significant. At the last count, one user had 154 systems. That number of systems using compressor technology would cost about US$7.5 million a year more to run than liquid nitrogen systems if the above difference is used in the calculation. Given the great differences in consumable cost per year, it is essential to perform an accurate comparison when selecting a thermal system.

Combining a liquid nitrogen boost with a mechanical system is another common mistake made by newcomers to the field. The mechanical systems are not capable of high-rates of change of temperature. When a liquid boost is used, the mechanical system is forced to change temperature faster than it can change itself and the mechanical system becomes a load to the liquid system and is detrimental to performance. It would be better to remove the compressor system totally; and just use liquid nitrogen. Furthermore, the mechanical system provides a large thermal load when heating the chamber because the heat exchanger coils are usually heated as well as cooled by the chamber air flow. Finally, the cost of such an adulterated system is exorbitant compared with the more efficient liquid system.

Maintenance costs differ so much between systems as to be remarkable! Examine each type of system for yourself and count the moving and stationary parts for a glimpse as to why this difference is so extreme. The mechanical systems have an excessive number of moving and stationary parts, each of which has its own failure modes and failure rates. The liquid systems have only two moving parts, a safety solenoid to stop liquid flow in the case of failure of the primary control valve and the primary flow control valve which can be either a solenoid or a proportional valve. There are no other moving parts in a liquid system except for perhaps a pressure regulator and/or a "keep cool" vent. The "keep cool" vents gas until liquid is present so that when liquid is commanded it is available immediately and rapid temperature cycling is possible after the long heat soak when no liquid was used. Gas would form in the lines if a "keep cool" were not present and there would be a long delay in receiving liquid nitrogen after opening the valves.

Reliability is obviously quite different in the two types of systems as is evident by the parts count as suggested above. It has been observed in many locations that one full-time mechanic is required for each seven compressor-based systems.

The size and weight of the two types of system is about 2:1 as shown in Table 9.2. Floor space is always at a premium in a manufacturing plant it seems and very heavy systems can sometimes only be placed on ground floors because of structural loading considerations.

Noise is a major consideration in most manufacturing plants.

Wearing ear protection continuously is a nuisance and is usually avoided by utilizing noise reduction techniques if possible. The mechanical systems are noisy at best. The liquid systems are nearly inaudible and can only be heard (in poorly designed systems) when solenoids snap open or closed.

The thermal range of the two types of system are vastly different. The liquid systems have been used to generate effective screens in only one precipitation and one detection cycle. *Many* effective screens have verified this use, but only a few have published on this subject. See [1] for more information.

Product ramp rates of 60 °C/min have been in use by the author in consulting situations and by seminar attendees for HALT and production HASS on many products since early 1989. The author has used rates up to 200 °C/min numerous times in the advanced seminar workshop and has not experienced rate-sensitive flaws. In his experience, which is admittedly not all-encompassing, no rate-sensitive flaw has ever been found. A few rate-sensitive flaws have been found and reported by seminar attendees, but, in all cases revealed to the author, a design change to allow higher rates in the production HASS would have reduced total production costs by a significant amount.

The liquid nitrogen or liquid carbon dioxide systems exhaust only clean gasses and so produce no pollution at all. The compressor systems do not produce significant pollution directly, but the cooling towers will collect pollution from the atmosphere. The water must then either be cleansed by some chemical and/or filtering process or the polluted water must be discharged into the sewer system for which there may be a charge. In one case, the pollution fee far exceeded the costs of electrical power.

The liquid carbon dioxide systems will not generate the very low temperatures used in HALT and HASS.

Both types of system have safety considerations. The liquid systems emit gasses which displace oxygen and, therefore, can be dangerous if used without proper venting. The gaseous discharge from the liquid systems should be piped to the outside of the building. The compressor systems are not dangerous unless there is a leak of the refrigerant. In that case, there may be safety problems to consider. Some refrigerants contribute to ozone layer pollution, but these have mostly been replaced by now, at least in the United States.

From the above comments on the systems under consideration it is seen that the liquid nitrogen systems are much less expensive to purchase and operate, more reliable, smaller, lighter, quieter, more efficient and non-polluting. The paradigm shift to the more advanced systems has been slow because of the profusion of mechanical refrigeration systems extant and the tradition of many decades of using them. Those still wanting to use the mechanical refrigeration systems are advised to seek out used systems which are available at a fraction of their original cost from many companies which have warehouses full of them because of the above discussed features.

9.5 DISTRIBUTED EXCITATION

When screening large systems such as large racks of electronics or maybe even a complete aircraft, for example, it may be beneficial to use distributed inputs. An aircraft undergoing HALT is shown in Figure 9.6. These tests are called structural tests in the aircraft business, but the philosophy is the same as HALT. The goal is to find the weak links in accelerated time. "Wiffle Trees" or a series of levers is used to distribute the loads on the wings, etc. to simulate a desired load distribution, such as flying through turbulence. Ever-increasing loads are applied until failure occurs, identifying a weak link or demonstrating that the structure is sufficiently strong to warrant advancing to production. These tests sometimes become very dramatic and may be heard from great distances when major structures fail! The tests are, of course, intended to cause failure.

Another example of a distributed vibration input would be to place a large rack of equipment on a shaker table and additionally to place small vibrators on the top of the rack. Such a system is shown in Figure 9.7, published with the permission of Array Technology, Boulder, CO. See [3] for more details on the use of this system. The rack weighed 1400lb and was excited to 20 GRMS all axes measured on the product and the product ramp rate was

Figure 9.7 Addition of vibration on the top of a rack (Courtesy of Array Technology Corporation)

20 °C/min also measured on the product. This is an example of a state-of-the-art system in 1991. The author founded the company that built the system.

In the example above, the use of distributed excitation can allow the upper and lower parts of the system to have the same or different vibration levels as desired. The six axes of vibration can also be tuned to a slight extent. That is, the vibrators on the top of the rack could all be pointed in the x direction in order to enhance the x vibration more than the y and z directions. At the same time, the base excitation could be tuned to be the same or different.

The thermal environments could also be tuned by using ducting from one thermal system to various parts of the unit or by using different thermal systems to excite various parts of the system to different temperatures or to different ramp rates. With distributed systems, almost anything can be accomplished in this regard.

Another example, used several times by the author, is to excite

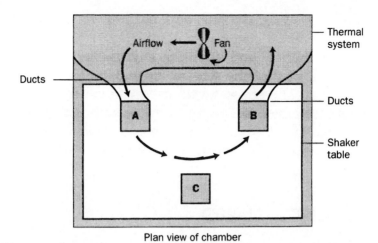

Figure 9.8 Chamber tuned for different vibration and temperature profiles at each position

different parts of a system, perhaps consisting of several boxes, to different vibration levels by regulating the air pressure to vibrators on the shaker table to different levels. This assumes the use of a pneumatically excited shaker table. An example of this type is illustrated in Figure 9.8 and is taken from an actual production HASS system. The figure is a plan view of a vibration/thermal system. The setup used distributed thermal excitation; that is, ducting to two of the boxes and a baffle in front of one box, to cycle the boxes to different levels and at different ramp rates. The temperature ranges, ramp rates and vibration levels were as shown in Table 9.3. The vibrators near box "B" were set to run at 100%, the vibrators near box "C" were set to run at about 60% and the vibrators near box "A" were set to about 20%, resulting in the Gs RMS as shown. The thermal ranges and ramp rates were

Table 9.3 Illustration of distributed excitation

Location	Vibration (GRMS)	Thermal range (°C)	Product ramp rate (°C/min)
A	5	180/–60	60
B	15	170/–50	52
C	10	160/–40	44

established by using ducting and baffling to get the values illustrated.

Tuning the thermal and vibration environments as discussed allows optimization of the screens for maximum efficiency and minimum cost. The stress levels used can relate to the in-use environments being different or to the operational and destruct levels being quite different. Other stresses such as voltage can also be distributed in order to optimize screens.

9.6 GETTING STARTED

It is normal to have to justify the purchase of new equipment in order to buy it. It is also necessary to have the new generation of equipment in order to obtain the data needed for justification. This conundrum presents an interesting problem for the company wanting to join the leaders in screening. A natural notion on the part of the beginner is to search the available literature for data, but the leaders are generally withholding the data because they wish to maintain whatever lead they have. A second step could be to rent or lease-to-purchase the equipment. Small bench-top models are ideal for the beginner because the units are relatively inexpensive and will accommodate a product of about $1ft^3$. The small units are suitable for use during production for troubleshooting failed units which will usually have to be performed at vibration and temperature in order to put the flaw into a detectable state. Comments on justifying costs to management are discussed in Chapter 10.

9.7 AUTOMATED HASS

The concept of fully automated HASS occurred to the author in 1984, but the concept was too far ahead of the industry to introduce then and the basic equipment for HASS was still just an idea. When the first HASS equipment was brought to fruition in 1991, the time for announcement of the AutoHASS™ was at hand and a patent was submitted in 1992. The patent was granted in 1996. The concept is to use a conveyor for the motion of the units to be screened and to move them through sequential chambers which are hot, cold, hot, etc. This motion of the product to be tested from chamber to chamber allows the chambers to stay at a constant temperature, saving an estimated 80% of the thermal energy usually used. Each chamber contains an all-axis shaker and loading of the pallet into the chamber and attachment of it to the shaker is fully automated as is the attachment of signal and power lines. Smart pallets are used so that mixed products can be screened in the same system with each type given its optimum temperature range, temperature rate of change and vibration level and appropriate diagnostics applied. It is anticipated that a screen in each station will last only 30–60 s, so throughput will be extremely high. This equipment is intended for companies screening very large numbers of units per day such as disk drives, computers, engine control modules, printers, etc. QualMark Corporation was assigned the patent rights to this invention. None have been built as of this writing.

9.8 SUMMARY AND CONCLUSIONS

From the comments on the thermal systems under consideration, it is seen that the liquid systems are less expensive to purchase and operate, more reliable, smaller, lighter, quieter, more efficient and less polluting. The "classical" equipment was designed to comply with the MIL-SPECs that came about during and after World War II. These specifications and design techniques are to a large degree no longer used and for good reasons. Therefore, there is a plethora of used "classical systems" available at deep discount prices. The paradigm shift to the more advanced systems has been slow because of the profusion of mechanical refrigeration systems throughout the industry and the tradition of many decades of using them. Those still wanting to use the mechanical refrigeration systems are advised to seek out used systems which are available at a mere fraction of their original cost. In the author's opinion, the equipment is not even worth the discount price, and he would not even take the classical equipment if it were free because of the gross reduction in efficiency when they are used. This applies to cost and to the ability to expose defects in design and processes.

The all-axis shakers available are vastly superior to the electrodynamic and mechanical shakers for all of the reasons cited before. As in compressor-type thermal chambers, there are many electrodynamic shakers collecting dust in warehouses and so these systems can be purchased at deep discounts if one is needed to comply with some specification.

The all-axis shakers and high-rate, broad-range thermal chambers are an integral part of the method and a few of the systems available were specifically designed to do HALT and HASS. If one has the classical systems, it is suggested that equipment designed for HALT and HASS be acquired. If this is not possible for whatever reason, then it is suggested that the equipment on hand be used with the HALT and HASS techniques as well as can be accomplished to prove that the methods work and then, with some experience to justify the purchase, acquire the equipment designed to do the job.

HALT and HASS are techniques that utilize the maximum possible time compression via high vibration levels using all-axis shakers, high rates of change of temperature and broad ranges of temperature. Other stresses are not considered in this chapter, but are frequently very important and have not been addressed because they are usually very product-specific and not easily amenable to general discussion.

REFERENCES

1. Hobbs, G. K., Triaxial vibration screening – An effective tool, *IES /ESSEH*, San Jose, CA, 21-25 September 1981.
2. Caruso, H. and Szymkowiak, E., A clarification of the shock/vibration equivalence in MIL-STD 810D/E, *Journal of the IES* , pp. 28–31, September–October 1989.
3. Hopf, A, Highly accelerated life test for design and process improvement, *1993 Proceedings of the IES*.

CHAPTER 10
Management Aspects of HALT and HASS

10.1 INTRODUCTION

This chapter covers various aspects of HALT and HASS which are of major concern to management. Subjects covered are: Proof of concept, the use of the methods by venture capitalists, evaluating a company for purchase, evaluation of competing companies, evaluation of HALT and HASS programs of other companies, outsourcing, some commonly made mistakes and a summary of the book.

10.2 PROOF OF CONCEPT

When first learning of HALT and HASS techniques, many people are very excited, are sold on the techniques, have learned enough to proceed on their own and want to start performing them right away. However, it will usually be necessary to present a business case for the investment in training, equipment, etc. The complete bill for all of these is large enough to get any manager's attention. If the manager has been educated on HALT and HASS, he will approve immediately. If the manager has not been educated on HALT and HASS, he will probably ask for proof that "the techniques will work on our products" and "How much will we save?" The answers to these questions are that the methods work on almost anything if applied properly and ROIs range above 1000:1. However, since the manager does not know this, some method of proving that the methods will work on the product built at your company must be found. This section will address that subject in five different ways: training, trial in-house with new equipment, calculated savings, trial in-house with existing equipment and a trial case at an outside laboratory. Several of these approaches will not work well as will be discussed.

10.2.1 Training: The Optimum Way

The fastest, most efficient and least expensive way to implement HALT and HASS is to have an in-house seminar on the subject and have appropriate staff educated in the methods in a fast, efficient and consistent way and to start using common terminology and concepts. The cost of an in-house seminar is truly insignificant if one appreciates what HALT and HASS can do for the company. Since this may not yet be known by managers in the company, it may be necessary to have some "scouts" go to an open seminar and appraise the content and its appropriateness for the company's situation and then have an in-house seminar. If this cannot be accomplished for whatever reasons, then the task is much more formidable and is outlined later.

After the "scouts" return and suggest in-house training, arrange for the manager(s) to attend a management overview of HALT and HASS. This could be part of an in-house seminar or could be separate. After the seminar, the manager(s) will probably not ask for proof that the techniques work and will direct you to proceed. First mission accomplished! The next things to do follow in order.

1. Have an in-house seminar for the design engineers, the production engineers, program management, quality and reliability engineers, purchasing, and a few others who may participate in the efforts. It has been found that if the education is done at the beginning, the whole program progresses much more rapidly and smoothly and therefore saves much more money. If

the education is not properly done in a timely fashion, then considerable amounts of money and time will be spent before things get under way, if they ever do. Initial education will pay back many, many times what is spent.

2. Buy the appropriate equipment to perform the HALT and HASS on your products. Note that a second vibration/thermal cycle system will be required for failure analysis and corrective action determination. A good way to start is to buy a small system for HALT and then use it for failure analysis during production. Do not try to use the same equipment for production HASS and failure analysis as it will create a bottleneck and will either stop production HASS or will stop failure analysis. Both are critical to a successful program and should be given the funding that they deserve. The ROI on the whole effort will be so high that to even bother to calculate it is a waste of time. To quote Gary Anderson, Vice President, Advanced Manufacturing, Storage Technology Corporation, Louisville, CO, "We do not have to justify buying HALT/HASS equipment because the ROI is so high that to waste time on the calculations is pointless". This is the voice of experience talking. Those without experience may not believe the statement and will spend vast amounts of effort performing a literature search and bench marking in an attempt to determine an ROI to be expected. This effort is doomed to be inaccurate because the best case studies are not published due to the advantages of the techniques; namely a very substantial technological and financial lead. At best, one could determine a lower bound on expected gains using what is found in the literature.

3. As the products are fielded, calculate the field failure rates from service data and compare with previous similar products. Publish the results within the company so that other programs or divisions also get the benefit of your experience so that they can proceed to perform their own HALT and HASS. It is very important to communicate the results within your own company because the results sound too good to be true when told by a seminar instructor or someone who has something to sell to you, such as shakers and chambers. When told by a company

employee based on the tests performed on the company's own products, the results tend to be believed.

This approach has been called "the leap of faith" by some and so it is. It is also the cheapest, fastest and best way to go. For some, however, the leap causes just too much apprehension and so there are four more ways to proceed that are less profitable.

10.2.2 Trial In-House with New Equipment: The Suboptimum Way

Without a demonstration on your own products, some management, particularly if they have not been to a seminar on HALT and HASS, will not approve funding for the program since they see it as a cost driver and also a schedule driver. It is indeed a driver, but *down on both counts!* This will not be apparent to the uninitiated in most cases. A plan in this case could be:

1. Send a few design engineers to an open seminar to get an education on HALT and HASS. There is more to it than found in the few published papers, so do not scrimp here.

2. Lease or rent equipment as necessary to run trial HALTs in your plant. This is superior to going to an outside lab because you will need support from failure analysis, production, spare parts, test gear, design engineers, etc. This is all readily available at the home plant, but not so at a remote site. You may want the help of a consultant for a few days or weeks during the first few efforts. Leasing or renting will generally require a commitment of about 60%–70% of the system purchase price because of the facilities costs and that presents a major stumbling block prior to management's buy-in to the techniques. The total cost of purchasing and installing is about 135%–145% of the system purchase price because of installation costs. Therefore, leasing or renting may amount to 50% of simply buying the equipment and installing it.

3. Once success has been proven, write up the results and distribute them to the appropriate staff to obtain approval for

going ahead as above in the optimum way. Note that some money and time will have been expended getting to the point where one could have started if the most optimum way had been selected in the first place.

This approach frequently fails due to lack of education and experience and so is not suggested. In addition, a substantial investment is required to finance what may be a failed attempt.

10.2.3 Calculated Savings: The Financial Estimate Approach

When one first learns of HALT and HASS, it is natural to want to try them on one's own product. Management will view this effort as an addition to all of the usual design qualification tests and so will see the effort as costing additional time and money. It is then necessary to discuss all of the aspects of the effort and consider the time and money spent on each activity and to try to justify the overall effort as a time and cost saver.

One of the first steps in convincing management to go ahead is to gather as much information on successful HALT programs as can be done. It has been usual in the past for most companies to treat this information as company private because to release it for general consumption would be to give away some company advantage. Bearing this in mind, one can look for successes from other industries, noting that Pareto charts from various industries are quite similar so that one need not restrict the search to a very narrow field. Once one has gleaned all of the available information from the available literature, one can sort the data and try to estimate the field failures that could be avoided in one's own product. Then an estimate of the costs of running HALT and HASS considering the purchase of a combined chamber for vibration and thermal cycling, test equipment, personnel, consumables and products used up in HALT and Proof of HASS is done. There are some cost savings items including the reduction of engineering development time, reduced warranty, reduced sustaining engineering, the benefits of increased customer satisfaction and also the effect of an earlier market release of a mature product. Additionally, there will be some losses due to fewer maintenance agreements and the sales of fewer spares.

It is readily apparent that many of the estimates used will have to be rough guesses and management is almost sure to argue about whether any given item is a saving or a loss. An example is design cost. Management will probably argue that the costs of design and development will be increased due to the additional tests to be run, the overdesign costs and the prototype units needed for HALT and the verification HALT. It takes one experienced in HALT and HASS to make all of the estimates and have them be believable and it is assumed that that experience does not exist in the company or the company would not be going through the exercise of calculating the savings. Based on the estimates that must of necessity be made, at this stage of experience, management will probably not place much credence in the estimates. This is not unusual at all and management will frequently reject the analysis as not correct. Another outcome is that the estimate based on available literature on poorly done programs may not justify going ahead with a HALT and HASS program. Now what can be done?

10.2.4 Trial In-House with Existing Equipment: Use What You Have

Send a few design engineers to an open seminar to get an education on HALT and HASS. Then use the existing equipment that is available which is usually the MIL-SPEC type of equipment not well suited to HALT or HASS. This type of equipment can uncover some gross design and process problems, but the equipment designed to perform HALT and HASS is far superior and will be better for a demonstration to management. Either way, it is best to do a HALT on a product that has known problems so that one does not have to fight the battle of convincing designers and others that the problems found are truly relevant. If the faults found are known to have caused field problems, then there is not much room for argument. If one uses a brand new product, the designers and others will argue that the product failed because it was tested out of specifications and so failure would be expected! For this reason, it is strongly suggested that an existing product with known problems, the more the better, be run through your first trial HALT. This is very true with any of the approaches mentioned herein and is very important.

The equipment on hand may not be adequate to expose enough opportunities for improvement to justify going ahead with a HALT and HASS program, so this approach may fail. There is one more way to proceed which may work and it is described in the next section.

10.2.5 Trial Out-of-House: Use an Outside Lab

If one desires to do a HALT with proper equipment designed to do HALT, then it may be necessary to go to a laboratory which has such equipment and also has staff knowledgeable in the methods. Be very careful here as there are many well-meaning testers who claim to know what HALT is and how to do it, but in reality do not have the skills or equipment to really do a meaningful HALT. Please note that one does not send a product to a HALT laboratory and get a report back. One must send design engineers, technicians, spares and monitoring test equipment to the laboratory. The laboratory will only assist in the running of the test and will supply the necessary stressing equipment. This is a very interactive test requiring your company's participation and your design engineers should have had some training in the HALT *and* HASS techniques. It is a major omission to try to skip this very important step, but many engineers have done just that.

After the test series of fault discovery and temporary corrective action on a product with known problems, design and process changes should be made, new hardware fabricated and then a verification HALT run to ascertain that the problems found have been fixed and that no more new problems exist. Most likely, several iterations will be required to reach the fundamental limit of the product.

If the trial HALT results in proof that the techniques found relevant problems on the product used (which it certainly will if done properly and there is something to be found), then it is time to run a HALT on a new product and to make some real progress instead of just demonstrating that the methods do really work on *your* product line. This next series of tests is sometimes best run at the laboratory used before so as to avoid buying the stressing equipment until management and the engineering staff are completely convinced that HALT is the thing to do. When they are finally convinced, one can then begin as in the optimum way. Note again that substantial time and money will have been spent before getting to the optimum way. Any way that you do it, education must be accomplished at some point and at the very beginning is the correct place to accomplish it.

A plan of action might be as outlined below.

1. Send a few design engineers to an open seminar to get an education on HALT and HASS. Going to an outside laboratory without knowing what is going to be done and why it works is far less than optimum since it always looks easy when you see an expert perform in any field, but it is much more difficult to do yourself and/or to explain it. There is an additional problem and that is that there are many laboratories offering "HALT services" when they do not have a good command of the methodology and may not have proper equipment either. There are millions to be saved using HALT and HASS, so learning how to do them properly just makes good sense. The author has seen many companies avoid training and then expend millions of dollars in a futile attempt to accomplish something that is really very simple, but fail due to ignorance which could have been cured for a few thousand dollars.

2. Arrange to have HALT performed on one or more products at a laboratory that has an intimate knowledge of the latest techniques and also has the appropriate equipment. Look for an all-axis broad-band shaker and a very high-rate of change thermal system. It is essential to have a success on your first attempt as you may not get a second chance if you fail on the first one. A way to succeed in proving the techniques on your products is to select a product for the trial which has known problems from field failures. Now if you find the problems, there will not be

any argument or doubt as to the relevance of the failures. If you were to select a new product which has had no field failures, perhaps because it has not been fielded, then inexperienced staff at your company would probably dismiss the failures as being caused by the overstress conditions used in HALT and would refuse to change anything, so no progress is made. This is the norm for this situation. In order to get around the problem of acceptance of the results and at the same time make some progress, it is suggested that one take along one new product and one old product with known problems so that at least some useful knowledge is gained to help the new product and, at the same time, the old product provides proof that the techniques are finding relevant defects.

3. Having shown that the techniques work on your products, one can now proceed to the optimum way above after having spent substantial funds and time to arrive at what could have been a successful starting point.

10.2.6 Conclusions on Proof of Concept

The HALT and HASS techniques are very simple once understood and have a body of established science behind them, but somehow the methods are not readily accepted by the novice who does not understand time compression. Proof that the techniques will "work on our products made in our plant" seems to be the necessary proof before in-house effort is supported by engineering and management. Without their collective approval and belief in the methods, progress will be very slow at best, if at all. Of all the five approaches outlined above, the first and the last have been found to work well with the first suggested as being the most cost effective and the last being less cost effective and slower to occur. The other approaches have been discussed only because many companies try them, usually without success. Education about HALT and HASS is crucial to success and is strongly suggested at the earliest possible moment for the staff disciplines mentioned above. It seems that there is some critical number and/or mix of people that need to accept the new paradigm before a total corporate commitment to the HALT and HASS methods can be accomplished. This number has been found to be 15–20 people of mixed positions within the company on a broad basis average. However, as few as two have accomplished success when they had full control of product design. As many as 1200 failed to succeed when they did not have product design control. This example refers to the US military in the mid 1980s when 1200 seminar attendees over a two-year period could not succeed in putting HALT and HASS into place even though they were personally sold on them.

Once a commitment is made to accept HALT and HASS, progress is rapid and cost savings start to accumulate very quickly.

Justifying HALT and HASS to management is sometimes a trying, difficult and expensive task. The steps suggested as successful have been found to be successful in nearly all cases in the last three decades. An attempt at HALT and HASS using unskilled staff and inappropriate equipment has usually failed to demonstrate the techniques successfully and has hardened the position of those in doubt to the point that it is then almost impossible to convince them any time in the near future because they have tried it and it did not work. This is not the place to try to economize.

An in-house seminar on the subject is usually much less expensive than fighting the misconception battles, one misconception at a time, in micro-seminars. Be very selective in picking an instructor and look for one who has the education, skill and experience to do the subject justice. An appropriate instructor can usually teach the techniques to the point of at least convincing the skeptics that the techniques are worth trying. Given a halfway successful trial, the job of convincing management is usually accomplished at this point and the process can proceed.

10.3 OTHER USES OF HALT AND HASS

10.3.1 Venture Capital Funding

Several venture capital firms now use HALT to evaluate the probable reliability of the products of a company being considered for the investment of capital. The author has observed this several times and even knows of a venture capital firm that *demands* HALT on a firm's products before providing funding to the company.

The way that a venture capital firm uses the process is to have an independent laboratory run a HALT on some or all of the products of the company being considered for funding in order to evaluate the potential success of the products in the market-place. This is done because the venture capitalists recognize that reliability as well as functionality and features are very important to the success of most products. A rugged product probably means a reliable product. High reliability will not necessarily lead to high profits, but low reliability will eventually lead to low profits, so the establishment of large margins is a necessary prerequisite for knowledgeable investors.

10.3.2 Acquisition of a Company and due Diligence

In one case of which the author is aware, the techniques of HALT were used to evaluate the robustness of the products of a company being considered for acquisition. The evaluation showed that the products had a few weak links and that the margins could be improved immensely by removal of the few weaknesses. Based on this, and other evaluations, of course, the target company was acquired and the opportunities for improvements shown to be possible by HALT were made resulting in very large financial margin increases due to reduced warranty costs and increased sales. That is, the company was purchased based on the financial margins extant upon evaluation and actual profits were based on the company after the opportunities for improvement were in place. The purchase was made at a price less than the true worth of the (improved) company.

10.3.3 Evaluation of Competing Vendors

Since the acceptance of the HALT and HASS techniques has spread to so many companies and product lines, the use of HALT to qualify vendor's products has also become very popular. If a particular company is striving to select the best subassembly for their product, a good way to do so is to compare the candidate product's robustness using HALT. In this approach, subassemblies, such as power supplies, for example, are obtained from various sources and then a HALT is run using all appropriate stresses. The upper and lower operating and destruct levels are then obtained for all candidate subassemblies and the candidates ranked from best to worst. This ranking is somewhat subjective since a given product may be outstanding except for one or a few weak links. However, once the weak links have been determined, the manufacturer may be willing and able to increase the design margins quite readily and thereby obtain an extremely robust design. Working with the vendors to ruggedize their products may be necessary in some cases and in others there will be an obvious best choice that is already far superior to all other candidates with no changes whatsoever. This candidate may well be (and usually is) the best choice from a reliability standpoint.

Not only is reliability important to most companies, but so also is cost. The best candidate may be the least expensive since the manufacturer may have used HALT and HASS on the products and therefore obtained the cost reductions possible with the techniques. Sometimes, however, the best product from a reliability standpoint is not the least expensive and may even be the most expensive. Now a more careful look into the total costs of a product, including the cost of the purchase of the subassembly, the integration cost, the test cost, the warranty cost and any other relevant costs such as the cost of not offering the best, or at least an entirely satisfactory, product in the face of stiff competition, is necessary. Loss of market share due to dissatisfied customers is also a necessary consideration. The most expensive subassembly may, and frequently does, turn out to be the least expensive when all the costs are calculated, or estimated in the case of market share loss. The choice of a subassembly based solely on the lowest cost may not be optimal in terms of overall cost and often results in disproportionate costs due to failures in the field use environment.

HALT will also find production flaws which will give a good indication of process control by the builder and will give an

EVALUATION OF OTHER HALT, HASS AND HASA PROGRAMS

indication of the probable field failure rate of the vendor's products due to inadequate quality control.

The author uses HALT results (from the manufacturer's tests) to select products for his own purchase. Examples from the last year when a new home was purchased include, clothes washer and dryer, trash compactor, cook top range, microwave range, refrigerator, dishwasher, phone system, fire detection system, garage door openers, florescent lights, burglar alarm system, motion sensors, sprinkler system, heating system, air filtering systems and air conditioning systems. Other items, not for the home, selected because of the large design margins included a global positioning system (GPS) for primary navigation during instrument flight for his airplane and a portable GPS for his automobiles.

10.4 EVALUATION OF OTHER HALT, HASS AND HASA PROGRAMS

There are numerous occasions when a company may want to evaluate the HALT, HASS and HASA programs of some other business that could be a subcontractor, a subsidiary or maybe even a company being considered for a takeover. Individuals may also want to evaluate potential manufacturers of items to be purchased for personal use. All of these situations will be discussed as one, because the approach is the same regardless of the relationship. Evaluation requires complete data as to what was performed, what the product is, the end-use environments and the transportation and storage environments as well.

Critical items that should be present are listed below. One should examine the procedures and equipment used in HALT. Specifically, the items for which to look include the following as a minimum;

10.4.1 HALT

1. Appropriate stresses for the product should be used. The stresses should be capable of precipitating the flaw types likely

to be present and should be related to the intended use environment. This does not mean simulate but to effectively screen for flaws that the environment would precipitate.

2. All-axis broad-band random vibration excitation should be used even if it is not present in the field. An exception is when no vibration-sensitive flaws are expected and also when vibration would not help detection, as in modulated excitation. A flaw driven to a hard failure by a chemical reaction might fit this description.

3. High-rate, broad-range thermal cycling should be used even if it is not present in the field. An exception is when no temperature- or rate of change of temperature-sensitive flaws are expected and also when temperature would not help detection, as in modulated excitation. A flaw driven to a hard failure by electromigration might fit this description.

4. Combined stresses should be used as appropriate. The bare minima are vibration, temperature and voltage (if appropriate).

5. Opportunities for improvement should be exercised up to the fundamental limit of the technology. This evaluation will require some investigation in depth since it is not easy to discern when the FTL is reached unless one is very familiar with the product, its manufacturing processes and all environments to which the product will be subjected.

6. Modulated excitation with verified high-test coverage monitoring should be used at appropriate intervals during HALT. *Appropriate monitoring is absolutely mandatory.*

7. Improvements should all be verified by re-HALT.

8. Focusing on failure modes and mechanisms should be done. Focusing on stress levels used will lead to missed opportunities for improvement.

9. Large margins on all expected in-use environments, including shipping, should be attained.

10. Fixtures used should transmit stresses to the product, not isolate it.

EVALUATIONI OF OTHER HALT, HASS AND HASA PROGRAMS

11. Vigorous failure analysis should be in place followed by a rigorous corrective action program. This is also *essential*.
12. Action should be taken on all weaknesses discovered.

The author has met some people who think that stressing until failure occurs and then doing nothing to improve the product is HALT. It is not! The improvements are what HALT is supposed to accomplish and, without the improvements, there is no progress toward reliability improvement.

10.4.2 HASS and HASA

1. Precipitation and detection screens should be used if precipitation above the operational limit is possible.
2. Verified high-test coverage monitoring should be used at least during detection. Partial monitoring during precipitation may be possible. *Appropriate monitoring is absolutely mandatory.*
3. Modulated excitation during detection screens will increase the detection tremendously.
4. All-axis broad-band random vibration should be carried out even if it is not present in the field.
5. High-rate, broad-range thermal cycling should be carried out even if it is not present in the field.
6. All appropriate field environments should be applied concurrently.
7. Safety of HASS should be performed. *This is absolutely essential.*
8. HASS optimization should be performed. This is very highly recommended and will reduce costs to the minimum possible commensurate with effective screens.
9. Re-HALT at appropriate intervals should be performed and margins must be maintained, otherwise the Safety of HASS will be invalidated. If margins cannot be maintained, re-do Safety of HASS and HASS optimization.
10. Statistical process control should be in place if it is suitable.

11. Vigorous failure analysis should be in place followed by a rigorous corrective action program. This is also *essential*.
12. Fixtures used should transmit stresses to the product, not isolate the stresses from the product.

If all of the items above have been done satisfactorily, then it is a safe assumption that the HALT, HASS and HASA programs are being properly carried out. If any items have been omitted, then it is a safe assumption that reliability is not what it could be. The program never ends as far as HASS is concerned, although it may progress to HASA if little is found in a properly performed HASS.

10.5 OUTSOURCING

Some comments on outsourcing are appropriate at this point. Outsourcing means to have another company make parts or assemblies for your own company. The desire here is usually to make more profit by reducing cost via economy of scale. That is a good idea, but there are some subtle points that are repeatedly overlooked. If the vendor is fully capable of doing a satisfactory job of assembly and closed loop HASS; that is, with precipitation, detection, failure analysis, corrective action and verification of corrective action, then the situation can work well. However, if the vendor does not have the capability or simply does not do a complete HASS including the five above steps, then the situation is almost certainly not going to lead to success. It can very likely lead to going out of business for the recipient of the products built by the outsource firm!

Among other problems, the various failure mode distributions, which were moved as far as possible from the normal operating range in HALT, will start moving toward the normal operating range, leaving less margin and invalidating the Safety of HASS that should have been done before starting production screening. This motion of the distributions is obvious from the second theorem of thermodynamics which says, "A system will always go to a lower organizational state unless something is done to change that state." When the margins decline, potentially "good" hardware can be seriously degraded by the screens, resulting in early wearout in the field. *This implies early life wearout failures of the entire population!* This situation appears to be very common [1], although most companies will not readily admit it.

The author has consulted to several companies which had begun outsourcing expecting to obtain cost savings. What actually occurred was grossly increased costs and vastly lower reliability due to movement of the distribution. This definitely is a very poor trend to let continue.

Another very serious problem occurs in the above mentioned scenario. It is that the systems manufacturing company, the customer of the outsource, incurs substantial costs in the troubleshooting and repairing of the circuit boards and also in warranty costs. The repair and warranty costs are usually not counted as the cost of manufacturing the circuit boards, as they should be, but they are assigned to some other account, such as sustaining engineering. In one case of which the author is aware but did not participate as a consultant, the cost of the in-house staff needed to repair the circuit boards manufactured out-of-house exceeded the cost necessary to manufacture them (including HASS in its entirety) when they were made in-house. Cost accounting was not correctly done and the costs of repair and warranty were assigned to sustaining engineering. Management incorrectly concluded that a large and expensive staff should not be necessary since most of the manufacturing was hired out to the outside source and so many of the in-house staff were laid off as a result. Then there were not enough qualified personnel to troubleshoot the hardware and to repair it. Field reliability was plummeting and costs were exploding when the last information from this large multinational corporation was received by the author. This situation is probably common.

The cure for the problem is very simple. One entity has to do all five steps of precipitation, detection, failure analysis, corrective action and verification of corrective action. This can only be done successfully in one place by one entity and so this also implies that either the systems manufacturer has to do everything; i.e. not outsource, or the outside manufacturer has to do all of them. Attempting to split the responsibility will result in a very poor situation that will be frustrating for both parties as well as not being satisfactory. This has been observed several times in the last few years. Failure analysis, corrective action and process control are difficult to do well at best, but separating the functions spatially and temporally results in substantially less than satisfactory results.

10.6 PITFALLS TO AVOID IN HALT AND HASS

Much has been written herein on what to do, but what *not* to do should also be discussed in order to enforce the positive. Some of the most commonly occurring mistakes are listed below and the reason why each is a mistake is then discussed briefly. For newcomers to the techniques, it is suggested that one keep this list nearby and refer to it once a month with the following question in mind, "Am I doing any of these?" If so, read the appropriate chapter(s) again. The mistakes are not necessarily listed in order of importance and all can lead to major technical and financial problems which can be easily avoided if the correct techniques and equipment are used.

10.6.1 Common Mistakes

1. *Attempting HALT and HASS without education on the technology.* Education is fundamental to success in many scientific fields including this one. Without a basic understanding of the principles and techniques, the person attempting HALT and HASS can be expected to make all of the mistakes listed below.

2. *Not performing Safety of HASS.* This is one of the worst possible mistakes that can be made. HASS is quite capable of reducing the usable field life of good hardware if the stresses chosen are too high for the design. If HALT has been properly done, then this rarely occurs. If HALT has not been done at all or if the margins attained are not high enough, then field life reduction may occur. The only way to reasonably prove that the screens are safe to use in production is to run Safety of HASS. If this is

not done, then the field failures will determine whether the HASS has been done properly or not. Clearly, this wait-and-see approach is not a good way to find out whether the screens are too intense because, by the time it is recognized that the screens are too intense, substantial numbers of reduced-life units will have been shipped. *Safety of HASS is mandatory for disaster prevention.*

3. *Not monitoring with high coverage during stimulation.* This is also one of the worst possible mistakes that can be made and usually leads to a substantial financial loss. What usually happens in this situation is that flaws will be precipitated but not detected and then the customer will have failures due to precipitated defects soon after delivery, leading to a very unhappy customer and many field returns. It is suggested that if one cannot monitor the product to at least a modest extent during stimulation with the various stresses, then the whole idea of HALT and HASS should just be abandoned. HALT and HASS (as well as classical screening and testing) without good coverage is less than worthless! Coverage can be determined and improved significantly by performing Software HALT (Chapter 8).

4. *Not using accelerated stress conditions.* If accelerated stress conditions are not used, then there will be no time compression. If only singular stresses are used, then the acceleration factor will be less than 1; that is, slower than in the real-world conditions which are almost always combined stresses. Without using accelerated stresses, it will take years to mature a product. This has been the result of classical validation testing as used by the auto companies and of qualification testing as used by the military contractors. Success-run testing is not suitable in today's accelerated product development scenario.

5. *Not improving the product to the fundamental limit of the technology.* When a weakness is found in HALT, the situation presents an opportunity for improvement. If advantage is not taken of the opportunity, then progress toward a failure-free

product will have been lost. It is generally found that almost everything found in HALT will show up in the field sooner or later. During HALT is a perfect time to improve the product as much as possible. Later, the fixes will be very expensive and the HALT investigator will appear very stupid indeed when the same failure modes are discovered in fielded hardware. One does not design to the expected HALT environments. HALT only identifies the weaknesses very quickly. It is generally very easy to improve the product to the point where it is more robust than it needs to be without spending undue funds in doing so. Since this is usually the case, it is foolish not to gain the very large margins that lead directly to high reliability and cost little or nothing to obtain. Everything found does not have to be improved, it just needs to be considered and those improvements that will improve field reliability or reduce cost are the ones that need to be done. It is better to err on the conservative side than to miss an opportunity. A large multinational company found that an average cost of missing an opportunity in HALT usually resulted in US$10,000,000 in field failures before a fix could be implemented.

6. *Not using HALT and HASS unless your competition is openly admitting or advertising doing so.* Many companies that are successfully using the methods will not admit to doing so as they have a technical and financial advantage with the use of the methods and it is to their benefit to remain quiet. Do not assume that a particular company is not using the techniques just because they will not admit to doing so. The author has performed seminars and consulting for at least 100 companies that will not admit to using HALT and HASS, nor can the author say that he has worked for the companies as required by non-disclosure agreements.

7. *Needing to verify that HALT and HASS will work on one's own products before starting a program.* This seems to be human nature and, therefore, occurs very frequently, many times because some of the results that have been reported seem far too good to be true. The methods are extremely versatile and have been used on hundreds of types of products. If one can

PITFALLS TO AVOID IN HALT AND HASS   205

figure out what failure mode is sought and how to stimulate the mode to occur and then detect the fact that a precipitated defect is present, then one can perform HALT and HASS. If one cannot figure out what is sought, one can just use the whole gamut of stresses and find out what shows up and fix those modes that seem to be relevant. Focus on the failure mode and not on the stress type or level which exposes the flaws. To focus on the stress type or level will lead to the conclusion that one should not fix many relevant flaws that should, in reality, be fixed.

8. *Requiring uniformity.* The large vibration tables designed for HALT and HASS generally have large observed variations in spectrum and overall level as one moves accelerometers around the table surface. This is not intentional by the designers, it just comes out that way. Chapter 6 discusses why uniformity is not necessarily required for excellent screens if HALT has been properly done. Buying one of the tables specifically designed for HALT and HASS and then mapping it out and trying to obtain uniformity will probably slow the introduction of use of the table by a large time factor during which time one could have been making progress in HALT or HASS. Safety of HASS and HASS optimization will determine whether the uniformity is good enough, whatever it happens to be.

9. *Missing the fact that the lowest and highest frequency modes of your product must be stimulated in order to perform a good HALT or HASS.* In general, all significant modes in all six directions must be excited to a reasonable level in order to obtain the desired effect. In addition, one must be able to monitor the product under test and to be able to detect any anomaly that may occur. This means that a broad-band all-axis excitation system that has little or no emitted magnetic field should be used. In specific cases, such as in simulating a vehicle passing over speed bumps or some other narrow band event, this comment does not apply and single-axis hydraulic shakers may be the optimum choice. The new broad band vibration system can also perform these stimulations.

10. *Using fixturing that does not transmit the stress to the product under test.* This is a lost cause because sufficient levels of the stress never reach the product. Three examples are: (a) using a vibration fixture that will not transmit the frequencies associated with critical modes of vibration of the product under test or isolates the mid and high ranges; (b) using a thermal fixture that does not transmit the conditioned air to the product such that the product can be rapidly changed in temperature over a broad range; (c) using electrical overstress and having some circuitry such as the lightning arrestor circuitry bleed off the high voltage before it gets to the internal circuits. If the stress does not get to the product, then nothing has been accomplished. The author has seen all three above examples and more in the field over the last thirty years.

11. *Using inappropriate vibration and thermal equipment.* Companies frequently find themselves becoming interested in the methods after acquisition of the equipment to do the old "MIL-SPEC" type of tests which are conceptually totally different than the accelerated philosophy. HALT and HASS are intended to break the product under test if it has a weakness, whereas the old MIL-SPEC tests were intended to *simulate* the environment accurately one stress at a time. Furthermore, the tests were often tailored for passing by clipping all stress levels three standard deviations above the GRMS level and maybe even notching out troublesome frequencies. Most classical test equipment cannot even get close to simulating the real environments, much less stimulating to time compression levels. The obsolete equipment can still be used for a low-stress HALT and HASS program to the point of demonstrating that the techniques work on the products produced by the company before the purchase of the equipment specifically designed for HALT and HASS. The proof of concept testing described in Chapter 10 may be useful as a means of proving that the techniques work on a specific product line to the point of purchasing new equipment.

12. *Not using simultaneous excitations of appropriate stresses.* This is not only technically much less effective than combined excita-

tions, but is also much more expensive since it requires more test equipment to monitor the product under test. In some comparisons, specifically in the seminar workshops over a three-month period, 100% of the defects detected using modulated excitations could not be detected at all with single excitations. Combined excitation is a quantum leap ahead of single excitations in effectiveness and is several times more effective in terms of cost due to reduced instrumentation and floor space required.

13. *Not using modulated excitation during detection.* Experience has shown that modulated six-axis vibration combined with slow temperature changes has exposed many flaws that could not be found any other way. Modern HALT and HASS equipment will easily do the modulation and it increases detection efficiency by at least a factor of 10 or more in many cases. In the seminar workshop, it has been repeatedly demonstrated that patent defects could not be found until the modulated excitation was done. Many times, 100% of the patent defects cannot be found without it.

14. *Not performing re-HALT.* The strength distributions mentioned in Chapter 2 and elsewhere in this book will be pushed as far as possible from the in-use environments during HALT. Screens will then be picked, proven to be safe and optimization done. If the strength distributions subsequently move toward the center, as one is assured that they will by the second law of thermodynamics, then Safety of HASS and HASS optimization will both be nullified. In this case, HASS can take an unacceptable amount of life from the products, leaving them unable to make it through a normal lifetime without failing in a wearout mode. The author has seen several companies make the mistake of assuming that everything would remain the same when it did not and never would. The results were screens of intensities that were too high for the product with its reduced margins, resulting in many early field failures. In two of these cases, essentially the entire population failed during the warranty period. It does not get much worse than that! Keep those margins up with re-HALT.

15. *Performing proof of concept on a mature product.* The point of proof of concept is to demonstrate that the methods work on your products. If a mature product is subjected to HALT, nothing should be found. This defeats the test and will lead to the incorrect conclusion that HALT does not work. It is suggested that one take a known problem product to the proof of concept demonstration so that there is something to find and the process can be proven to work. This will convince all concerned that the methods very quickly expose weaknesses in the product.

16. *Concentrating on short-term costs.* This has been the classical failing of American management over the last few decades. One must get beyond thinking in terms of shipments this quarter regardless of reliability, resulting from the necessary shortcuts taken to obtain the "numbers" and focusing on short-term outputs at the expense of the long-term attainment of goals. Shareholders and owners are interested in the long term profits and stability of the company. Management attitudes should reflect this interest. HALT costs money upfront in the design phase, the returns come later in design, production and field service. The emphasis should be on long-term profitability that can be gained with a comprehensive HALT and HASS program.

17. *Not optimizing the HASS.* The task of optimizing HASS is relatively simple and straightforward. It usually takes a few weeks to accomplish in high-production programs. Cost savings have amounted to 50%–80% of the pre-optimization costs in many cases. Considering the costs of HASS during a high-volume production program, it only makes sense to reduce the cost as much as possible while also proving that the HASS regimen exposes all flaws that are present. If the initially chosen HASS regimen is not quite enough to precipitate all latent defects, then those remaining will occur in early field life, perhaps during the warranty period.

18. *Outsourcing without closed-loop corrective action.* Many companies outsource, anticipating economy of scale. Precipitation,

detection, failure analysis, corrective action and corrective action verification should all be done at one location. The outsource manufacturer's plant is the correct place to do them.

19. *Terminating HALT and HASS activities after initial successes.* One company terminated HALT and HASS activities after attaining the position of world leader in their field. The second law of thermodynamics states that this position cannot be retained without utilizing a vigorous HALT and HASS program, perhaps using HASA. Designers will go back to their old ways if the product does not face HALT before production. This has been observed several times by the author. It is a real shame to attain the lead and then to regress to the back of the pack.

10.7 USING MIL-HDBK 217 OR ITS DERIVATIVES

Some well-known documents such as MIL-HDBK 217 and derivatives of it treat all flaws as being precipitated by temperature alone, which is completely erroneous. See [2] for many discussions of why the handbook is incorrect. Some component problems can be found by burn-in and that is why the component manufacturers use burn-in so successfully. Many companies that are assembling above the component level are discontinuing burn-in because it has not been found to be cost effective. In order to test this thesis, it is suggested that HASS be performed and then, on the same products, burn-in performed. It is generally found that very little is discovered in the burn-in, and when the cost of performing burn-in is considered, the usual decision is to terminate it. Most of what can be found by burn-in can be found in a proper HASS thermal cycle. As a matter of general interest, it is noted in passing that the Arrhenius equation has been incorrectly used to describe any number of failure modes which do not follow the equation at all. MIL-HDBK 217 was a prime example of the rampant misuse of the equation. See [2] for a series of discussions of the subject. See also [3] for a discussion of the independence of failure rate on temperature for microelectronic devices. In the author's opinion,

MIL-HDBK 217 should be immediately placed in the shredder and all concepts therefrom simultaneously placed in one's mental trash can. If this statement upsets the reader, it is suggested that the reader obtain both references above and thoroughly study them several times. MIL-HDBK 217 will go down in history as one of the biggest impediments to progress ever promulgated on the technical community. The ISO series may be in the same league, but may have some value for companies with little quality organization.

10.8 SUMMARY

The HALT and HASS techniques are very simple once understood and have a body of established science behind them, but somehow the methods are not readily accepted by the novice who does not understand time compression. Proof that the techniques will "work on our products made in our plant" seems to be the necessary step before in-house effort is supported by engineering and management. Without their collective approval and belief in the methods, progress will be very slow at best, if at all. The cost reductions attainable *over the first year alone* will greatly exceed the cost of implementation. Education about HALT and HASS is crucial to success and is strongly suggested at the earliest possible moment for the staff disciplines mentioned before. It seems that there is some critical number of people that need to accept the new standard before a total corporate commitment to the HALT and HASS methods can be successful. Terminating the HALT and HASS techniques will inevitably lead to a regression in reliability. Continuing HALT and HASA seems to be a good compromise between expense and results.

HALT has many uses other than the ruggedization of products. Among these are:

1. Venture capitalists are using it to evaluate the potential for a company to make money prior to funding the company.
2. Interested acquirers are using it to perform due diligence to determine what the potential profits from a company's

products could be if all opportunities for improvement are exercised.

3. Companies use it to evaluate various vendor's products in order to select the most reliable one. Individuals use the supplier's HALT data to select products for their own use.

A checklist has been supplied for use when evaluating another's HALT and HASS programs, including HASA.

Many common mistakes have been observed over many years of teaching HALT and HASS and also observed in performing consulting on a very diverse set of products. The most important of those mistakes has been elaborated upon in the hope that the reader can avoid the common pitfalls. It is suggested again that the reader review the mistakes on occasion to determine whether he has fallen into one of the common mistakes, and, if so, re-read the appropriate chapter.

Properly done HALT will reduce design time, reduce design cost, reduce sustaining engineering costs, reduce warranty cost and will also improve customer satisfaction and reliability substantially. When using HALT properly there will be little or no reliability growth during production because it will all be gained during design. Vastly improved margins leading to increased market share will make the company producing the products more profitable than it would have been otherwise. Experience shows that HALT will make the field failure rate of a product much less, but HALT gives no measure of what the failure rate will actually be. When HALT and HASS are properly done, it seems to be a waste of money to even try to estimate field failure rates. It would be much more cost effective to use the money to perform HALT on another product instead.

Properly done HASS will reduce the time to discover manufacturing problems, reduce sustaining engineering via corrective action, reduce screening equipment costs by orders of magnitude, reduce consumables such as power and liquid nitrogen, reduce the floor space required for production, reduce personnel and reduce the warranty costs. The results will be improved field reliability and customer satisfaction which will result in increased market share and profitability over the long term.

The two techniques used in harmony will lead to products that cost less to produce and seldom, if ever, fail in normal field use. The products usually become obsolete before any failures occur.

Continued use of the HALT and HASS methods will eventually lead to a situation where the company's designers have become "product smart" and are designing products wherein little or nothing that needs to be improved is exposed in HALT. Also, the production department will be producing hardware that seldom fails in HASS and so HASA will be in place and will expose little that needs to be fixed. In that case, the company can be said to have arrived at the desired quality and reliability. World-class quality and reliability are really quite easy to accomplish. HALT and HASS are two tools which make the attainment of these goals financially and technically viable.

It seems that there is some critical number of people that need to accept the methods before a total corporate commitment to the HALT and HASS methods can be successful. Once a commitment is made to accept HALT and HASS, progress is rapid and cost savings start to accumulate very quickly. Once the HALT and HASS approach has begun, you will wonder why it took you so long to accept them. The tools are ready, you just need to avail yourself of them.

REFERENCES AND NOTES

1. Schenkelberg, F., Product reliability and outsourcing the electronic assembly process, *Proceedings of the 1999 Accelerated Reliability Technology Symposium*, Hobbs Engineering Corporation, San Jose, CA, 17–21 May 1999.
2. *Quality and Reliability Engineering International*, **6**, (4), September–October 1990, John Wiley & Sons. There are many excellent articles on failure prediction methodology in this issue, which is must reading for anyone using MIL-HDBK 217-type methods.
3. Hakim, E. B., Microelectronic reliability/temperature independence, *Quality and Reliability Engineering International*, **7**, pp. 215–220.
4. O'Connor, P. D. T., *The Practice of Engineering Management*, John Wiley & Sons (1994).

APPENDIX
Glossary of terms

Acceleration – The rate of change of velocity expressed in gravitational units where 9.8 m/s*2 or 32.2 ft/s*2 is one g.

Acceleration spectral density – A plot of the amplitude versus frequency with the ordinate of g^2/Hz and the abscissa of Hertz. This is a frequency-domain representation of the vibration.

Accelerated Test – The use of stresses above the field environment so that the desired effect takes place in less time (time compression).

Accelerometer – A transducer with an output proportional to acceleration.

Ambient environment – Room conditions of temperature, humidity and vibration.

Amplitude – The magnitude of a variable. Can be RMS or peak.

Accumulated Fatigue Damage Factor (AFDF) – Uses Miner's Criterion to sum fatigue damage. Cannot be used to compare vibration systems since only a single-degree-of-freedom model is used in the analysis. The single degree of freedom as used here means how many degrees of freedom the mathematical model had, not how many axes of shaking is generated by the shaker. See any text on introduction to vibrations for clarification

Arrhenius equation – This equation describes the reaction rate of some chemical reactions, gaseous diffusion rates, migration effects and precious little else. The equation is the basis of MIL-HDBK 217 which purports, quite erroneously, to predict failure rates of electronic systems. Many reliability prediction techniques are based on this equation which does not predict correctly more than a few percentages of the failures today, and for this reason, the prediction techniques are badly flawed to the point of being less than worthless.

Acceleration Spectral Density (ASD) – Average acceleration power in each narrow band of an analysis of vibration. Same as power spectral density (PSD).

Audit – A test (HASS) on a portion of the total lot. The sample percentage may be very small, less than 1%.

Broad band – Refers to vibration over a reasonably broad-band, usually greater than 2000 Hz. Some currently available systems can supply meaningful vibration over the band from 0-10000 Hz.

Burn-in – Temperatures well above the normal operational temperatures are used to accelerate the process of fault precipitation as described by the Arrehenius equation. Burn-in is largely used successfully only by electronic component manufacturers today since it does not precipitate most flaws that exist in today's electronics above the level of components.

Could Not Verify problem (CNV) – Other names are: Could not duplicate, No Defects found and Retest OK.

Component – A non-repairable item, e.g. capacitor, resistor.

Corrective action – To implement a change intended to eliminate the source of the flaw in future production or in a design after fully understanding the root cause. Should be verified.

Crest factor – The ratio of peak to RMS value of a waveform. Also called Peak to RMS ratio.

Critical damping – That damping value which will result in zero displacement in minimum time.

Device Defect Tracking (DDT) – A database which helps track a problem from its discovery, to root cause, to implementation, to lessons learned, so that it is never repeated.

Degrees of freedom – In mechanical engineering, the number of directions in which an object is free to move, generally six. Motion described as side to side would be only one degree of freedom: sideways.

Design limit – The operational or destruct limit of a product. The limit beyond which it will not operate correctly or is broken, respectively.

Design margin – The difference between the capability of the product and the expected field environment.

Design ruggedization – The process of finding the design weak links and fixing them so that the design is very robust; that is, has capability far beyond that required in the expected field environment. Note that the field environment is rarely, if ever, accurately known, therefore large design margins are generally in order.

Destruct limit – The stress level above (or below) which the product suffers irreversible harm to the point that it will not function properly when returned to normal stress levels.

Detection screen – A combination of stress application combined with testing of some sort to detect that a failure of some kind has occurred. Has the ability to see a patent defect after it has been processed through a precipitation screen.

Distributed excitation – Multiple input stimuli at various locations. Forces may include vibration, voltage, thermal and others. Generally useful on larger structures such as cabinets and larger assemblies.

Destructive physical analysis (DPA) – Examples are SEM, dissection.

Device under test (DUT)

Electromagnetic interference (EMI)

Exponential acceleration – An acceleration of results which is related to the stimulus in an exponential way.

Failure analysis – Determine the origin or root cause of the flaw.

Fatigue Life – The time under some operational conditions that a product is expected to survive. Can be rated in hours, miles, operations or by some other means.

Field environment – The environment(s) which exists in the normal use situation.

Flaw coverage – The total percentage of flaws which can be detected by the software and test technique in use.

Fluidic burn-in – The use of fluoroinerts to thermal cycle electronic equipments. Generally does not include vibration and so has poor precipitation and detection.

Frequency margining – Using varying frequencies of input stimulus to find the range of operational frequencies.

Gs in a root mean square (GRMS) – A G is the acceleration due to gravity and is 386 in/s^2, 32.2 ft/s^2 or 9.8 m/s^2. This parameter alone means absolutely nothing all by itself! The frequency content, in some meaningful way such as the PSD and the probability density are needed in order to calculate the effect on the specimen. GRMS is the square root of the area under the PSD curve.

Highly Accelerated Life Tests (HALT) – Done to ruggedize the product and obtain large margins over the expected in-use conditions. Uses all stresses which can cause relevant failures. Stresses are not limited to field levels or stresses.

Hard failure – A failure which is detectable when stresses are removed.

Harmonic – An integer multiple of some frequency. Only objects which follow the wave equation have harmonics in terms of vibration. These would include vibrating strings such as on a violin, a vibrating drum head, and vibrating air masses such as an organ pipe and other wind or stringed instruments. Objects such as a printed circuit board follow an equation like a plate, which does not exhibit harmonic behavior. Multiple modes do exist, but the frequencies are not harmonically related.

Highly Accelerated Stress Audit (HASA) – This is HASS in a sample form. Very cost effective if quality and design margins are very good.

Highly Accelerated Stress Screens (HASS) – Screens in which stresses much higher than expected in the field are used in order to gain time compression.

Highly Accelerated Stress Test (HAST) – Stresses are: temperature, humidity and atmospheric pressure. Condensed water is not desired, only high water vapor pressure to permeate seals and cracks as well as to drive chemical reactions. Corrosive materials are sometimes used to accelerate reactions.

Infant mortality – A failure which occurs early in life, usually due to a workmanship defect.

Laminar air flow – Air flow where the streamlines are reasonably straight and do not cross or mix. Turbulent flow is where the streamlines are mixed up and flow is usually unstable. Heat transfer is much higher when turbulent flows exist.

Latent – Undeveloped, dormant and usually undetectable by testing the product's performance.

Life test – Exposing the product to the field environments until end of life occurs or some other criteria has been satisfied, e.g. running an automobile up to 300 000 miles without any wearout modes occurring (would not count tires and other replaceable items).

Linear – Having a straight-line relationship on linear paper between two variables such as input and output. Most electronic systems are not linear in terms of vibration response to input. This makes them very difficult, if not impossible, to analyze accurately.

Liquid Nitrogen (LN^2) – Used as a coolant in rapid thermal cycling systems. The temperature of the nitrogen is approximately $-160\,°C$.

Manufacturing defect – A flaw introduced by an improper process or technique during manufacturing.

Miner's Criterion – The relationship which expresses the fatigue damage done in terms of a summation of a number of cycles at various stress levels. See any advanced strength-of-materials text for details.

Modulated excitation – Consists of slowly varying temperature and rapidly varying all-axis vibration excitation. Sweeps through the two space defined by the temperature and vibration ranges used. An excellent detection screen when used with monitoring and perhaps other stresses as well. It is also called a "search pattern". Not limited to only vibration and temperature in general.

Monitoring – Checking the performance behavior of a product, usually during stressing.

Mean time between failures (MTBF) – This is when the product is operated at assumed known operating conditions. This property has nothing to do with wearout! MTBF can be related to the height of the bathtub curve in the lower (flat) portion of the curve after infant mortality and before wearout. Many state-of-the-art methods do not even speak of MTBF because it is not relevant in many ways and is only measurable by utilizing very long and costly tests on many products under normal operational conditions or in the field environments. One assumption made in stipulating an MTBF is that of constant failure rate.

Narrow band vibration – Vibration of limited frequency range, e.g. from 500 Hz to 520 Hz.

NAVMAT P-9492 – The NAVY P-9492 document of May 1979. This document suggested random vibration and thermal cycling as a means to find latent defects in electronic equipment. This much was excellent. Unfortunately, the document has been followed as a MIL-SPEC, which it is not, to the detriment of many manufacturers. This document is only a guideline. The thermal profile in it is "NOT TO BE USED" to quote the document, and the vibration profile is only an example and does not include the critical low and high frequencies. This document has no value today except for historical purposes and is not recommended reading and certainly is not recommended as a screen tutorial document.

No defects found (NDF) – Synonymous with Could not duplicate, Retest OK and Could not verify.

Notching – The elimination or reduction of discrete frequencies from the vibration.

Non-Linear – A system which does not have a linear relationship between the input and the output. Electrical systems are usually very non-linear in vibration amplitude response.

Octave – A doubling of frequency such as from 100 Hz to 200 Hz.

Octave rule – A design rule wherein the modes of vibration are kept at least one octave apart starting with the most overall mode and going down to local modes as the frequency doubles. The reverse octave rule is where the innermost mode is the lowest and the most overall mode has the highest frequency. A more general design rule is to avoid resonances at the same, or nearly the same, frequency because magnification may occur; that is, very large accelerations may occur leading to very large stresses and probably to failure.

Original equipment manufacturer (OEM) – One who integrates a product supplied by another into his own product and then sells it.

Operational limit – The applied stress above which the product will not operate with complete functionality. Usually, going slightly above this stress will not damage the product, it will just not function above it. Reducing the stress will usually result in normal behavior.

Paradigm – A pattern, example, tradition or model. Paradigms die slowly and with great difficulty and usually take many years to change except among the leaders in a field. The three stages of a paradigm shift are: (1) Ridicule; (2) Denial; and finally (3) Acceptance as being obvious.

Parameter drift – The shift in a parameter over time such as the time required to perform a certain act or function. The shift could also be in some performance parameter such as bandwidth or frequency response. Some early life settling in is usually normal but, after that, changes are usually indicative of a problem.

Patent – Adjective used to describe something that is detectable under the correct conditions. Coverage is required in order for detection.

Periodic – Implies that it is repetitive and therefore its state can be predicted.

Proportional, Integral, Derivative (PID) – Parameters used to tune and control a simple closed loop system. Commonly used in thermal control systems.

Particle impact noise detection (PIND) – Detecting loose particles with an acoustical test. The shaker head has a built-in microphone that detects the noise of the bouncing particles.

Power cycling – Turning the power on and off.

Power spectral density (PSD) – A measure of the power of a random signal measured over some bandwidth. The dimensions are $(GRMS)^2/Hz$. The square root of the area under the PSD curve over some frequency range is the RMS value over the same range. The probability density is also needed in order to calculate fatigue damage.

Product ruggedization – The process of determining the limits of operation/destruct of a product and then improving the limits through design and/or process changes.

Peak Probability Density Function (PPDF) – Rain flow analysis. Measures only peaks and the number of occurrences. Omits small cycles as not of importance.

Precipitation – Changing some flaw in the product from latent (non-detectable) to patent (detectable).

Proof of screen – This is a two-part process. The first demonstrates that the screen does not remove too much life from a product and that the product is still suitable for shipment to a customer after repeated screens and the second demonstrates that the screen is effective. The former is essential and the later is not recommended. Currently called Proof of HASS.

Power spectral density (PSD) – A narrow band analysis of a broad-band signal. This measure is valid only if the signal is stationary and ergodic. This assumption is not true for the impact type of excitation and so the PSD is of little use in the evaluation of a transient signal. This analysis method is valid for traditional random vibration. Same as ASD.

Random vibration – Vibration where the signal is stationary and ergodic in time; that is, the signal at any one time has the same statistical properties as at any other time.

Rapid thermal cycling – Cycling the product at high-rates of change of the products under test.

Repetitive shock system – A system wherein the vibration is caused by a series of shocks.

Resonance – A condition of maximum response in terms of

frequency. Increasing or decreasing the frequency will reduce the response.

Reliability Growth Development Testing (RGDT) – A test in which normal operating stresses (assumed to be known, frequently erroneously) are used to test many products for many hours. When problems are found, corrective action is taken. These tests take a very long time because non-accelerated stresses are used. These tests are now obsolete compared with the HALT techniques which use time compression.

Root Mean Square (RMS) – The average intensity of all signals. When used for vibration, this measure is valid only when other meaningful quantities are known such as the PSD and the probability density of the signals. RMS is totally meaningless when used to describe vibration all by itself.

Root cause – Understanding the reason for the failure. Its true understanding is proven by being able to turn the problem on and off as the fix is inserted and then removed.

Retest OK (RTOK) – Synonymous with No defects found and Could not verify.

Seeded samples – Samples which have had intentional defects inserted.

Scanning Electron Microscope (SEM)

Signature analysis – Measuring the response of a product to signals of varying characteristics.

Six degrees of freedom vibration – Vibration which has acceleration in three translations and three rotations, all simultaneously. There are several subsets of this general type of vibration available from various manufacturers and all are substantially different.

Slow temperature slew – A situation where the rate of change of temperature is very slow such as 30°C/min. Currently, rates of change of more than 60°C/min on product are considered to be moderately fast.

Soft failure – A failure that is only detectable under specific conditions, but is not detectable under ambient conditions.

Spec limit – The (assumed to be accurately known) operational stress levels. Seldom, if ever, are the actual field levels accurately known.

Shock response spectrum (SRS) – This is the peak response of a single degree of freedom oscillator (linear spring, mass and linear damper system) to a transient where the peak response is plotted versus resonant frequency. An SRS can have any number of time domain signals that generate it. The SRS gives only the one-time peak response to a transient and gives no information as to fatigue damage. Only useful for a one-time transient such as squibs firing and if the dynamic system has reasonable damping so that substantial ringing does not occur.

Step stressing – Progressively increasing the stress levels when looking for design weaknesses.

Scanning Tunneling Microscope (STM)

Stress screen – Application of stresses in order to find design or process defects.

Stiffness – The ratio of force to the corresponding change in displacement. Usually non-linear in electrical systems of normal construction.

Stress plus Life (STRIFE) – A subset of HALT, conceived by Hewlett-Packard. Commonly used by Hewlett-Packard even if they are using HALT.

Test Analyze And Fix (TAAF) – A long test where normal operating environments are simulated and problems sought. No time compression is usually used.

Thermistor – A resistive element whose resistance decreases with increasing temperature. Limited linearity, operating temperature range and the lack of robustness preclude its use in HALT and HASS.

Thermocouple – The joining of two dissimilar metals producing a junction whose output is proportional to temperature. Used in HALT and HASS.

Tickle vibration – Low level all-axis impact excitation of varying levels and directions used for detection screens. Has developed into modulated excitation.

Time compression – When something occurs faster than in the normal field environment, usually due to stresses above the field environment.

Turbulent air flow – Air flow where the streamlines and velocity are not steady.

Voltage cycling – Turning the power on and off or varying the voltage to the product.

Voltage margining – Testing at varying voltages to determine the voltages at which normal operation can be expected. Usually temperature sensitive.

Index

acceleration signal 167, 168
acceleration transform 130, 133
acquisition of companies 195
all-axis vibration
   effectiveness 36, 88–9
   equipment 17–18, 166–74, 184
   see also modulated excitation;
     vibration
Anderson, Gary 187
Array Technology 142, 179
Arrhenius equation 15, 138
AST system 170, 171, 173
at-speed functional testing 151
automated fault injection (AFI) 99,
     153–7, 159–60
automated HASS 182–3
automated signal integrity testing 161–2

'bathtub curve' 3–4
Boeing 21, 22
boundary conditions 47
built in self test (BIST) 148
burn-in 209
   defined 138
   equipment 174
   flaw-stimulus relationships 140, 141

Cadillac Luxury Cars 72
Casagranian telescope 6, 7
'classical equipment' 165, 183, 206
clock variation 138
combined stressing 85, 88, 141–3, 206–7
   see also modulated excitation

company acquisition 195
competing vendor evaluation 195–7
compressor cooling *see* mechanical
     refrigeration
corrective action 24, 79
   verification of 24, 79
corrosion 139
costs
   cost-benefit relationships 54–6
   financial estimates 189–90
   savings 5, 7, 8, 66–9, 133
coverage
   defined 33
   importance of 87, 98–9, 103, 108, 203
   in software HALT 34, 149–50,
     157–60
CUBE^{TM} system 169, 170, 171, 172

defect levels, estimate of 157–60
design ruggedization 4, 121–4
   see also HALT
'design smart' 69–70, 79, 131, 211
design verification test (DVT) 66, 68
destruct limits 35, 36
detection
   defined 23, 78
   detection screens 6, 78–80
     selection of 87–9
   probability of detection (software
     HALT) 148–9
discriminators 37–8, 46, 93–4
distributed excitation 179–82
documentation 79

due diligence 195
DVT-100 154

Edson, Larry 22, 72
effectiveness of HASS 97–8
  proof of 107–12
electro static discharge (ESD) 139
electrodynamic shakers 169, 170, 171
electromagnetic interference (EMI) 139
electromigration 85, 198
environmental stress screening (ESS) 8, 12, 13
  see also HASS; screening
equipment 17–19, 165–84
  automated HASS 182–3
  'classical equipment' 165, 183, 206
  distributed excitation 179–82
  getting started 182, 187
  inappropriate equipment 206
  thermal equipment 174–9
  vibration equipment/systems 17–19, 166–74, 184
evaluation
  of competing vendors 195–7
  of other programs 197–200

failure
  analysis 23–4, 78
  'bathtub curve' 3–4
  distribution 2–3, 40–2
  fatigue failure 2, 3
  modes 9–12, 32–3, 40, 65–6
  non-relevant 52–3
  overload 2, 129
  patterns of 2–4
  physics of see physics of failure (PoF)
  soft and hard failure 92, 93
fatigue damage 2, 3, 104
  see also mechanical fatigue damage
fault coverage see coverage
fault injection 151, 152
  automated fault injection (AFI) 99, 153–7, 159–60
  manual fault injection 152–3
financial estimates 189–90
fixturing 47, 205–6
flaw–stimulus relationships 140–3
four corner tests 48–9

fragile elements, protection of 51–3
'fundamental limit of technology' 5, 39, 203–4

General Motors 72

Hakim, Ed 137
HALT 31–75
  becoming 'product smart' 69–70, 79, 131, 211
  comparison with classical approaches 25
  defined 1–2, 4, 31
  equipment required see equipment
  evaluation of other programs 197–9
  examples of successes 19–23, 28–9
  level at which to do HALT 56–7
  location of testing 58–9
  mean time between failures (MTBF) 2, 72, 74, 130
  number of units to test 60
  other uses for HALT 194–7
  overview 4–8
  paradigm change and 64–6
  personnel involved 57
  phenomena involved 15–17
  protection of fragile elements 51–3
  purposes of 8–12
  and reliability 1–2
  repeated HALT (RE-HALT) 53, 62–4, 207
  resistance to accepting 70–2
  step stress approach 49–51
  step stress intervals 53–4
  stimuli applied in 9–11, 35–49
  stopping HALT 54–6
  terminology 34–5
  time and cost savings see costs; time compression
  uses for tested units 60–1
  virtual HALT 130
  see also software HALT
hard failure 92, 93
HASA
  evaluation of other programs 199–200
  principles of 9, 26, 119–20
HASS 77–101
  automated HASS 182–3

comparison with classical approaches 25
defined 1–2, 8, 77
detection screens 6, 78–80
  selection of 87–9
development 25
discriminators 93–4
effectiveness 97–8
  proof of 107–12
equipment required *see* equipment
evaluation of other programs 199–200
examples of successes 28–9
  of HASS alone 82–3
fault coverage and resolution 98–9
on field returns 99
HASS tuning 96
margins of strength 34
objectives 80–1
optimization *see* HASS optimization
other uses 194–7
overview 4–8
phenomena involved 15–17
precipitation screens 6, 8, 36, 78–80
  selection of 85–7
probability of success 100–1
product response to stimuli 84–5
proof of *see* proof of HASS
purposes of 8–12
and reliability 1–2, 83–4
safety of *see* safety of HASS
screening process 78–80
setting up 94–5
stress magnitude selection 81–2
see also modulated excitation
HASS optimization 25, 113–20, 208
  process of 114–18
  see also HASA
HAST 107, 139
Hewlett–Packard 5, 11, 39, 55, 125, 204
highly accelerated life tests *see* HALT
highly accelerated stress audit *see* HASA
highly accelerated stress screens *see* HASS
highly accelerated stress test *see* HAST
Hobbs Engineering Corporation 170
humidity 139
hydraulic shakers 169, 170

In-house seminars 59, 186–8
in-house trials
  with existing equipment 190–1
  with new equipment 188–9
'infant mortalities' (failure of weak items) 3, 4
Institute of Environmental Sciences (IES) guidelines 13–14

Jedrzejewski, Dave 142

Leonard, Charles 21, 22
life, reduction by screening *see* safety of HASS
life tests 25
liquid carbon dioxide cooling 178
liquid nitrogen cooling 174–9
load 2, 3
lower destruct limit 35
lower destruct temperature limit (LDTL) 38
lower operational limit 34
lower operational temperature limit (LOTL) 38
lower voltage destruct limit (LVDL) 48
lower voltage operational limit (LVOL) 47

management aspects 185–212
  evaluation of HALT, HASS and HASA programs 197–200
  other uses of HALT and HASS 194–7
  outsourcing 200–2, 208
  pitfalls to avoid 202–9
  proof of concept 185–94
  using MIL-HDBK 217 or derivatives 209–10
manual fault injection 152–3
manufacturing yield 157, 158, 159
margins 32, 34, 42
  verification of 63–4
mean time between failures (MTBF) 2, 72, 74, 130
mechanical fatigue damage 15–16
  how HALT and HASS work 131–4
mechanical refrigeration 174–9
mechanical vibration systems 169, 170
microphonic noise signature 93–4

MIL-HDBK 217 15, 26–7, 209–10
MIL-SPEC tests 21, 84, 85, 165, 183, 206
Miner's criterion 15–16, 132
mode of vibration 84, 166–7
Modular™ vibration system 169, 170, 172–3
modulated excitation 89–93
  defined 37
  effectiveness of 37, 45, 93, 138, 206, 207

NAVMAT P-9492 12
Nortel 22

octave rule 139
OmniAxial vibration system 169, 170, 174
operational limits 34, 35, 36
operational vibration level (OVL) 46
optimization *see* HASS optimization
Otis Elevators 22
outsourcing 200–2, 208
overload failure 2, 129

particle impact noise detection (PIND) 140
physics of failure (PoF) 32, 129–46
  choice of stresses 137–9
  defined 129
  examples
    rate of change of temperature 135–7
    vibration 134–5
  mechanical fatigue damage and HALT/HASS 131–4
  types of failure 2–4, 129–30
  Venn diagrams 140–3
piezoelectric shakers 169, 170
pitfalls to avoid 202–9
power cycling 85, 138
power spectral density (PSD) 167
precipitation
  defined 23, 78
  precipitation screens 6, 8, 36, 78–80
  selection of 85–7
Premier™ vibration system 170, 173
probability density 167
'product smart' 69–70, 79, 131, 211
production screening *see* HASS
proof of concept 58, 59, 185–94, 207–8

proof of HASS 103–12
  proof of effectiveness 107–12
    see also HASS optimization
  proof of safety *see* safety of HASS
protection of fragile elements 51–3
Proteus Corporation 149, 154

Qual tests 5, 25, 68, 96, 105, 106
QualMark Corporation 170, 172, 183

random vibration 88, 166, 167, 171
refrigeration equipment 174–9
reliability 1–2, 83–4
reliability demonstration (Rel-Demo) tests 2, 20, 25, 106
repeatability 121, 125–7
repeated HALT (RE-HALT) 53
  importance 207
  verifying design capability 62
  verifying margins 63–4
resolution 98–9
  defined 33, 150
  in software HALT 150–1
return on investment (ROI) 56, 71, 186

S–N diagrams 16, 132–3
safety of HASS 17, 34
  compared to classical approaches 25
  defined 95
  importance of 202–3
  process involved 95–7, 104–7
sample size 5, 33
sampling 119
screen clearance 122
screening
  history of 12–15
  see also HASS
search pattern *see* modulated excitation
seeded samples 97–8, 108–12
seminars 59, 186–8
Seusy, Cliff 33
shipped defect level 157–60
short-termism 71, 208
show screens 62
signal integrity evaluation 161–2
sinusoidal vibration 167
six sigma approach 42
soft failure 92, 93

software HALT 6–7, 34, 147–63
  at-speed testing 151
  automated fault injection (AFI) 99, 153–7, 159–60
  automated signal integrity testing 161–2
  coverage 149–50
  detection 148–9
  field experience with customer use 152
  manual fault injection 152–3
  resolution 150–1
  shipped defect level in production 157–60
  simulated field experienc 152
  software fault insertion 152
  time domain reflectometry (TDR) 162
step stress approach 49–51
  intervals suggested 53–4
  when to stop 54–6
Storage Technology Corporation 82, 187
strength 2, 3
stresses, types used 9–11, 35–49
STRIFE 5
stuck-at fault coverage 151, 154, 157, 158, 159

Team Corporation 170
temperature/thermal cycling
  distributed excitation 180, 181–2
  effectiveness of 138
    combined with vibration 12–13, 45, 87–9, 141–3
    thermal cycling alone 13, 45
  equipment 174–9
  flaw-stimulus relationships 140–3
  HALT process 37–46
  rate of change 45–6, 138
    acceleration factor 135–7
  see also modulated excitation
Thermotron 170
tickle vibration 79, 88, 89, 100
time compression 5, 8, 32, 66–9, 130, 133, 203

time domain reflectometry (TDR) 162
training 186–8
trials 59
  in-house *see* in-house trials
  outside laboratory 191–3

uniformity 121–5, 126–7, 205
upper destruct limit 35
upper destruct temperature limit (UDTL) 38
upper operational limit 34
upper operational temperature limit (UOTL) 37–8
upper voltage destruct limit (UVDL) 48
upper voltage operational limit (UVOL) 47

Venn diagrams 140–3
venture capital funding 194–5
verification of corrective action 24, 79
vibration
  acceleration factor 134–5
  combined with temperature cycling 12–13, 45, 87–9, 141–3
  distributed excitation 179–82
  effects of 139
  equipment/systems used 17–19, 166–74, 184
  flaw-stimulus relationships 140–3
  HALT process 46–7
  random vibration 88, 166, 167, 171
  see also all-axis vibration; modulated excitation
vibration destruct level (VDL) 46
virtual HALT 130
voltage 47–8
  flaw-stimulus relationships 140
  voltage variation effects 138–9

wearout failure 3, 4
wiffle trees 179
Williams and Brown model 157

yield 157, 158, 159